Exploring non-human work in tourism

De Gruyter Studies in Tourism

Series editor
Jillian M. Rickly

Volume 5

Exploring non-human work in tourism

From beasts of burden to animal ambassadors

Edited by
Jillian M. Rickly and Carol Kline

DE GRUYTER

Despite careful production of our books, sometimes mistakes happen. We apologize for not having included the part title of Part 5 in the original print publication. This has been corrected.

ISBN 978-3-11-065975-7
E-ISBN (PDF) 978-3-11-066405-8
E-ISBN (EPUB) 978-3-11-066004-3
ISSN 2570-1657

Library of Congress Control Number: 2020949158

Bibliographic information published by the Deutsche Nationalbibliothek
The Deutsche Nationalbibliothek lists this publication in the Deutsche Nationalbibliografie; detailed bibliographic data are available on the Internet at http://dnb.dnb.de.

Typesetting: Integra Software Services Pvt. Ltd.
Printing and binding: CPI books GmbH, Leck

MIX
Papier aus verantwortungsvollen Quellen
FSC
www.fsc.org
FSC® C083411

www.degruyter.com

Acknowledgements

The editors would like to give a big "THANK YOU" to the contributors of this collection. When we first envisioned this book, it was a simple idea. We wanted to demonstrate to the field the need for further consideration of the work that animals perform in the tourism industry. The authors responded to this goal by going so much further than originally imagined. They have presented work that pushes the boundaries of tourism studies theoretically, implements new concepts, and provides strong cases that implore tourism scholars to initiate more critical investigations.

We must also send a special note of appreciation to our colleagues who acted as reviewers for the chapters in this book. They no doubt helped to strengthen the collection overall with their constructive feedback and sharp insights. In no particular order, we thank: Kellee Caton, Nitasha Sharma, Jane Bone, Kevin Markwell, Kristen Lamoureux, Lauren Duffy, Alana Dillette, Brent Lovelock, Kelly Bricker, Stefanie Benjamin, José-Carlos García-Rosell, and Bryan Grimwood.

https://doi.org/10.1515/9783110664058-202

Contents

Jillian M. Rickly, Carol Kline

1 Introduction: Working for the (hu)man in the tourism industry

An animal labor perspective – why now?

Tourism is an ideal context for the study of human-animal relations, as tourism facilitates numerous types of animal interactions. Indeed, "one would be hard pressed to find a destination where animals were not used in some capacity for tourism purposes – animals held in captivity, circuses, rodeos, racing, fighting, pursued in sports, and as workers" (Fennell, 2015, p. 27). The research focused on wildlife tourism (Shackley, 1996; Higginbottom, 2004; Newsome, et al. 2005; Lovelock, 2008; Rodger, Moore, & Newsome, 2009), as well as captive animals as attractions (Cohen & Fennell, 2016; Desmond, 1999; Frost, 2011; Yerbury, Boyd, Lloyd, & Brooks, 2017; Winter, 2020), has arguably garnered the most attention in the field, with scholars interested in tourists' motivations and experiences of such encounters. However, along with a broader renewed interest in the ethical concerns of tourism, known as the "moral turn" (Caton, 2012), researchers have begun interrogating the ethics of our engagements with animals in the tourism industry with fresh perspectives. Such work reconsiders the notion of human-animal relations with greater emphasis on animal sentience and welfare (Fennell, 2012; Carr & Broom, 2018; Carr & Young, 2018). In fact, Fennell (2015) identifies five distinct theoretical approaches to animal ethics that have important implications for human-animal interactions in tourism: welfarist, animal rights, utilitarianism, ecocentrism, and ecofeminism. Moreover, Winter (2020) observes several ethics frameworks in her analysis of 74 articles and several books related to animals and tourism, including those articulated by Fennell, as well as instrumentalism. Rather than recite each of these theoretical approaches here, we encourage the reader to continue on with the chapters that follow, as these frameworks are each covered in varying detail therein.

As a result of this moral turn and the rise of critical animal studies in the field, greater attention is now being given to the labor animals perform in the tourism industry. Indeed, much of the work animals do in tourism goes unnoticed. While the animals who work as performers might quickly come to mind, such as dolphin tricks, elephant rides, or falconry displays, there are many other types of labor that are supplementary to tourist attractions, including horse-pulled carts for transporting tourists, cat cafes, or fish pedicures at spa destinations. Further still, some highly valuable animal labor is mostly performed out of sight, such as sniffer dogs in security and customs settings. Can these animals be labelled as "employees" if they are not paid, asks Carr and Broom (2018) of animal "workers" in the tourism industry? When considering the work animals do, it is vital to maintain our attention to the fact that this "work" is interpreted through human eyes and for human

https://doi.org/10.1515/9783110664058-001

needs. As a result of such thinking, we must also ponder what "work" means to animals and if it can be conceptualized as a dialectical relationship to "leisure." This is a theme taken up by both Dashper (Chapter 2) and Carr (Chapter 3) in the following two chapters, respectively, as the authors navigate the theoretical implications of this construct and its positionality.

While acknowledging that work is a human construct, we aim with this book to extend a critical lens to interrogate the diverse forms of animal labor found in tourism. In doing so, we reveal the nuances of animals as entertainment, while also drawing attention to the value-added labor performed tangentially to tourism attractions, as well as the hidden labor that is performed outside of our view or simply easily overlooked.

Welfare and animal (workers') rights

While scholars are increasingly raising questions as to the ethics of human-animal interactions in tourism, this book treats the issue of ethics as secondary as it aims primarily to investigate the kinds of work that animals perform in tourism. Of course, such investigations often include discussions of ethics, as several authors in this collection explain (see Fennell, Chapter 15; Garcia-Rosell & Tallberg, Chapter 7; Lopez & Venegas, Chapter 5; Tully & Carr, Chapter 6; Vasilopoulou, Chapter 10; Venegas & Lopez, Chapter 4). Nevertheless, this book is far from an exhaustive analysis of the ethical dimensions of animal labor. Towards that goal, we would recommend David Fennell's (2012) *Tourism and Animal Ethics* and Neil Carr and Donald M. Broom's (2018) *Tourism and Animal Welfare*, as well as a number of edited collections, including Kevin Markwell's (2015) *Animals and Tourism: Understanding Diverse Relationships*, Neil Carr and Janette Young's (2018) *Wild Animals and Leisure: Rights and Wellbeing*, and Carol Kline's (2018a) *Animals, Food, and Tourism* and (2018b) *Tourism Experiences and Animal Consumption: Contested Values, Morality, & Ethics*. As a result of focusing foremost on animal labor, we engage less with the complex and nuanced ethical debates regarding the question of whether animals *should* be exploited within the tourism industry. While a worthwhile question in its own right, we know that animals *do*, in fact, work in the tourism industry. As such, that is where we begin this conversation. This starting point thus inspires a discussion of the relations of animal welfare, well-being, agency, and their rights as workers.

Animal welfare, according to Hill and Broom, is "the state of an animal as regards its attempts to cope with its environment" (2009, p. 532). As Fennell (2015, p. 30) notes, animal welfarists are concerned with the quality of animals' lives "not whether animals should be used by humans" (see also Bekoff & Nystrom, 2004). According to Bekoff and Nystrom (2004), an animal's quality of life relates to their basic needs of food, water, shelter, but also extends to their treatment by humans, that is, whether humans are fulfilling their obligations to animals in terms of keeping them from pain and suffering. Yet, these responsibilities are set against the benefits that humans derive from the use of animals such that human benefits can sometimes be deemed to outweigh the pain, suffering, and even death of animals. Assessments of animal welfare, then, consider distinct aspects of animal life, including their natural behaviors, health and physiology, and feelings and mental state (see Hewson, 2003), and often use a scalar measure (very poor to very good). While generic applications of these categories to all scenarios can be problematic, as animal lives and living conditions vary so drastically (see Kistler, 2004), Fennell (2015, p. 30) summarizes the "five freedoms" as a generally accepted measure of gaging animal welfare: freedom from hunger and thirst; freedom from discomfort; freedom from pain, injury and disease; freedom to express normal behaviors; and freedom from fear and distress. More specifically, Carr and Broom (2018, p. 17–19) identify a range of needs that are essential to welfare:

- To have sufficient oxygen and good quality air
- To carry out feeding to maintain themselves and grow
- To have good environmental water quality
- To maintain osmotic and other body fluid stability
- To receive appropriate sensory input
- To rest and sleep
- To exercise and have space for movement
- To have appropriate social interactions
- To avoid fear
- To explore
- To thermoregulate
- To carry out reproductive and maternal functions
- To maintain good health condition
- To perform maintenance and eliminatory behavior
- To avoid pain and injury

While a more subjective concept than welfare, some organizations that work with animals choose to also concern themselves with animal agency. A brief search on animal agency outside of tourism reveals a robust body of work addressing the topic with abundant tourism examples: race horses, captive whales, sled dogs, hunted animals, etc. (Dietz & York, 2015; Hribal, 2007; McFarland & Hediger, 2009). Indeed, Nance (2013) devotes an entire volume to the agency of elephants working in the circus. Additionally, within animal studies literatures, topics of agency, sentience, and moral standing are discussed extensively; for example, Carter and Charles (2013, p. 322) distinguish between primary agency and corporate agency, noting animals "act and their actions have consequences, they also resist conditions which they do not like and, in some circumstances, are able to change the conditions of their agency." Further, Špinka (2019) outlines four levels of agency: passive/reactive agency, action-driven agency, competence-building agency, and aspirational agency. The "essential core of the concept of an agent," argues Steward (2009, p. 221), can be characterized by "how we should go about deciding which animals might be thought of as falling under it." Dietz (2006, p. 45) examines the role of animals as workers, among other uses, suggesting:

> animal agency is not always at the service of the production of human well-being. Animals act in idiosyncratic ways that occasionally can be dangerous, vexing, and somewhat inefficient when they are viewed as a resource used for production. Thus, the history of human interactions with domestic animals has involved the invention and refinement of methods to exert control over the agency of domestics – sometimes gently, sometimes cruelly.

Hribal (2007, p. 105) outlines many forms of animal resistance, and asserts that "the labor and resistance of animals have influenced other members of human society – some of whom saw commonalities in their mutual struggles against such forms of exploitation." If agency and resistance among animals are of such import as to inspire humans to include them in justice movements, why then is this not examined within tourism?

Certainly the animal *welfare* body of literature within tourism is growing, however very little has been said to date about animal agency beyond the point that often animals have little to none. Tourism scholars agree that animals are historically treated as objects and resources (Burns, 2015; Fennell, 2012; Shani & Pizam, 2008) but there are more nuanced dynamics to examine that are related to all of the actors within a tourism experience or network. In Winter's (2020) synthesis of animal ethics literature, little mention of the qualities of animal agency arises, demonstrating we are only in the beginnings of exploring animals within tourism literature. Two notable exceptions addressing animal agency in tourism are Notzke (2019) who explores agency of horses within equestrian tourism, learning that horses act as stakeholders within the tour groups, playing different roles which influence their riders' self-actualization and psychological experience, and Dashper (2017, p. 207) who explains that treating horses as individual sentient beings shifts a "moral responsibility on humans to safeguard animal interests in human sport and leisure." Because tourism is ultimately an economic pursuit, animals as actors within tourism must comply with the systems in charge. Most, if not all, of the chapters in this volume speak to animals' lack of agency while engaged in tourism work, and some chapters begin to tease out the nuances within and towards animal agency (Garcia-Rosell & Tallberg, Chapter 7; Hoarau-Heemstra & Nazarova, Chapter 8; Speiran, Chapter 11).

Animal labor in the tourism industry

Attention to animal welfare and agency forces us to acknowledge that animals are currently not afforded absolute freedom to choose their "work." Rather, they have been trained, directed, and coerced to complete particular tasks. It is humans who see value in the actions animals can perform. While research attuned specifically to animals as workers is growing, there are clear hurdles we still must address in order to move the field forward. Birke and Hockenhull (2012, p. 2, *emphasis original*) observe,

> in practice many empirical studies of humans and animals focus more on humans *or* nonhumans rather than the *relationship* between them [. . .] there has been focus on the outcome of our relationships with other species, but much less on how relationships *work*, as a process, or an ongoing interaction between two or more sentient individuals.

Relatedly, Dietz & York (2015, p. 46) elaborates on a dominant type of relationship developed within a working context,

> resources become capital when they are organized to produce profit and growth. We then note that some types of resources – – humans and many non-human animals – – actively resist control by those organizing the production process. Thus to increase the efficiency of the production process, those seeking to use resources as capital have to find a way to either reduce their need for resources with agency or to reduce the resources' agency.

Thus, we aim with this collection to draw attention to one dimension of human-animal relationships in tourism, specifically working relationships. This is accomplished a number of ways, including examining the symbolic value of animals (Bueddefeld & Benbow, Chapter 12; Guo, Chapter 13; Lopez & Venegas, Chapter 5;), analyzing the value that animal labor creates (Hansen, Rickly, Halpern, & McCabe, Chapter 9; Tully & Carr, Chapter 6; Venegas & Lopez, Chapter 4; Vasilopoulou, Chapter 10), observing animal-to-animal labor (Speiran, Chapter 11), and embracing an animal-centric position (Carr, Chapter 3; Garcia-Rosell & Tallberg, Chapter 7; Hoarau-Heemstra & Nazarova, Chapter 8) when possible.

Following on from current efforts to conceptualize animal labor, we have found Urbanik's (2012) categories a useful starting point: animals for education (used in labs and zoos), service animals (for strength, transportation, protection, and guides), and animals for recreation (sources of entertainment in films, circuses, fights, etc.). Yet, in examining the work animals perform in the tourism industry, we find that they often cross these boundaries. Indeed, several chapters in this collection demonstrate that boundaries are quite porous. As Venegas and Lopez show, for example, the "zonkeys" of Tijuana are used for transportation (service) and photography (recreation). Similarly, in the chapters by Speiran and by Bueddefeld and Benbow, the authors elaborate on the ambassadorial roles of primates and polar bears, respectively, and the types of educational and recreational work that these animals perform. What's more, such an approach to understanding animal labor in tourism also reveals how value realized in one type of labor can lead to further laboring. For example, animals used for transportation (service) may become iconic to a destination, leading tourism operators to further utilize the animals for photography and tourist entertainment (recreation), but also increase their popularity as zoo animals (education).

Additionally, we find Coulter's (2016) categories of animal labor helpful for adding further depth and nuance to the types of work animals perform, for themselves as well as for humans:

- *Care work* encompasses the provision of physical and emotional care, including protection and nurturing.
- *Communication work* involves the use of shared language, including non-verbal signals.
- *Emotion work* captures the ways animals manage their emotional states, as well as bonds between themselves and others, both humans and non-humans.
- *Body work* is the physical labor animal bodies are engaged in.
- *Dirty work* addresses the deeming, degrading and/or undesirable work that animals are sometimes forced into.

Indeed, we can observe these categories throughout the chapters in this book. Care work and emotion work are examined among primates in sanctuaries (Speiran, Chapter 11), as well as care work and communication work between guide dogs and their owners (Hansen et al, Chapter 9). Communication work and body work are noted among sled dogs, horses, riders, and guides (Garcia-Rosell & Tallberg, Chapter 7; Hourau-Heemsta & Nazarova, Chapter 8), while body work and dirty work are also the focus of the two chapters on donkeys (Vasilopoulou, Chapter 10; Venegas & Lopez, Chapter 4) and the chapter about bullfighting (Lopez & Venegas, Chapter 5).

Considering the categories of animal labor delineated by both Urbanik (2012) and Coulter (2016), we have devised the following approaches to animal work in the tourism industry: performative work, value-added work, and hidden labor. Rather than expanding the number of categories, which is an infinite exercise in attempting to capture all possibilities, we maintain fewer with the aim to be broad and relational (Figure 1.1). These approaches highlight most importantly that work is oftentimes agile, flexible, and multifaceted. They have also been established with an effort to consider the human-animal relations that perpetuate animal labor in tourism contexts. As Higgin (2012, p. 85) explains, "Animals are not somehow 'out there' in nature, they are 'in here' with us. We encounter them within specific relations, in particular places." In this way, we situate animal labor within broader concepts of work, as the deeply inequitable structures of tourism mean that it is not only non-human animals that are marginalized in the labor they perform. Attention to the labor and to their industry value offers a glimpse into the economic justifications at present, which in turn opens up a discussion for future advocacy and change.

Further, in conceptualizing these approaches – performative work, value-added work, and hidden labor – we are acknowledging the specific expectations that accompany touristic animal encounters. Within these approaches come certain touristic motivations and expectations. Thus, using these categorical approaches highlights what have become socially accepted human-animal experiences, which lays bare human-centric desires over and above animal well-being.

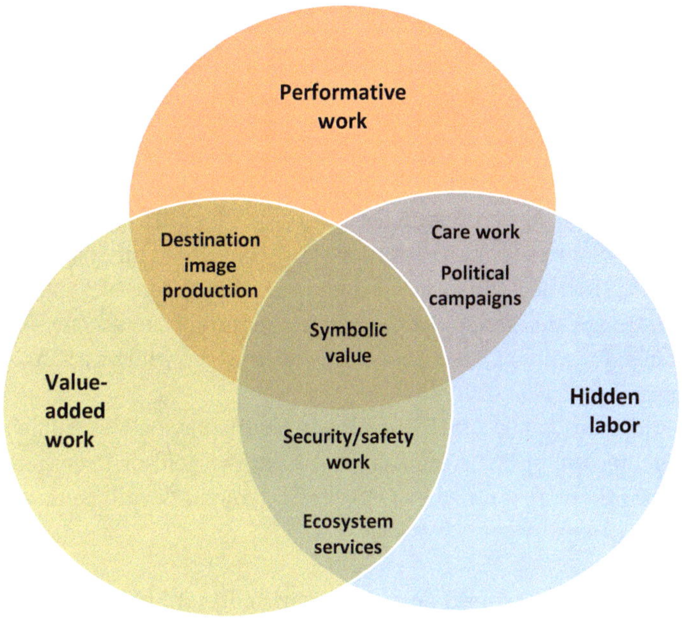

Figure 1.1: Approaches to animal work in tourism.

Performative work

When considering the work animals do in the tourism industry, it is often performances and entertainment that first come to mind. Indeed, Urbanik (2012) specifically mentions animals as tourist attractions within the category of recreational labor. Animals as attractions has a long history of tourism research (see Frost, 2011). Indeed, among the many ways that humans consume animals in tourism (Kline, 2018a; 2018b), visual consumption as objects of tourist gazes is particularly prominent. This includes animals in zoos, and other captive environments, where they may be simply gazed upon or trained to perform specific tasks or feats for viewers. However, "seeing animals seems not to be enough for many people; they wish to interact with them as well" (Carr & Broom, 2018, p. 32). As such, petting zoos and other attractions that bring humans into contact with animals extends visual consumption to satisfy other senses and emotions. In such settings, the animals are primarily valued for their performative work, even if that means sitting still and controlling any aggression while being stroked by humans.

Animal performances are not relegated to particular settings. Cohen (2009, p. 102), following Goffman's stages of social performances, distinguishes four types of tourism settings in which animals are the main attraction:

- *Fully natural settings* lack framing and restraints such that animals are wild and free to move. These include relatively large spaces of wilderness, jungle, desert, ocean, and so on.
- *Semi-natural settings* appear to be unbounded spaces in which the animals are unrestrained and therefore free to roam and move. Yet, they are highly managed and monitored spaces, such as national parks and some wildlife sanctuaries.
- *Semi-contrived settings* are simulations of the animals' natural habitats, such as zoos and aquariums. The animals are either captured from the wild or are from breeding programs.
- *Fully contrived settings* are constructed specifically for displaying the animals as performers. Whether the animals are captured from the wild or have been bred, they are more tamed as a result of intensive interactions with humans, including training practices.

While we might not initially think of wild animals in their natural habitats as performers, tourists do in fact bring an abundance of expectations about what they will see in terms of the animals' daily activities. In particular, the very expectation to see a wild animal in its natural habitat, and the ways this feeds into visitor satisfaction, demonstrates the often unrealistic expectations tourists place on animals. The highly visual nature of tourism experiences, particularly in relation to the circle of representation, which demonstrates the process by which destination images are continually (re)produced through both official promotions and user-generated content, is influential of tourists' expectations (Jenkins, 2003; Urry & Larsen, 2011). As such, tourism promotions and social media content that highlight tourists' encounters with wild animals establishes similarly high standards in terms of visibility, and therefore performance and entertainment value, to captive and domesticated animals. The actual encounters with animals on wildlife tours has direct consequences for tourists' satisfaction. As a result, reviews posted on social media platforms give incentive to tour operators to push further into natural habitats, as well as to bait and lure the animals out into the open where they are made visible to tourists (D'cruze et al., 2017; Knapp et al., 2013; Orams, 2002).

In these ways, wild animals in their natural habitats are expected to perform to similar standards of captive and domesticated animals. While there are limited opportunities for intimate encounters with wild animals in such settings, it is the semi-natural to contrived settings that afford those opportunities. Settings such as zoos, wildlife parks, and animal rescues offer not only closer proximity and stronger guarantees of observing animals doing their "biological" work, particularly as there are few (if any) places within the enclosures for the animals to hide. Moreover, such settings sometimes extend the types of work animals do to include entertainment.

Captive animals are tamed to facilitate handling by tourists and trained to perform tricks, notable cases include petting tigers at "sanctuaries" (Cohen, 2012) and swimming with dolphins in Australia (Mann et al., 2018).

Animals as performers are not limited to constructed environments. In destinations around the world, animals perform for tourists on streets, such as monkey shows or talking parrots. More commonly, still, animals trained for certain tasks in farm settings, such as herding dogs, are employed as both farm workers and performers for tourists as a means to economize traditional agricultural practices. Thus, as we see in the other approaches to understanding animal labor, animals as performers might also simultaneously be employed in value-added work or hidden labor as well.

Value-added work

Whereas some tourists might not so easily recognize "performing" animals as "working" *per se*, because it might appear, for example, that the animals at safari parks are "following their instincts" or dolphins in shows are even "having fun," Markwell (2015, p. 9) argues it is the beasts of burden are "most directly and clearly 'at work'" in tourism contexts. Beasts of burden are animals employed to pull or push heavy loads, such as carrying goods and assisting in crop planting/harvesting. In tourism, such animals are commonly involved in the provision of transportation along tourist transit routes; particularly, pack animals, such as donkeys, horses, and camels, often transport tourists and their luggage (see Cousquer & Allison, 2012; Markwell, 2015; Vasilopoulou, Chapter 10). Thus, in addition to entertainment performances, animals also work in tourism settings as part of the overall destination experience. While tourists rarely travel to see these animals specifically, their presence in the destination adds to the overall experience and may even authenticate the place as they represent part of collective social discourses and geographic imaginaries (Tully & Carr, 2020). For example, very few tourists travel to Santorini, Greece, just to see its donkeys. Nevertheless, the donkeys have become a notable and novel part of the tourist experience as they "transport tourists and/or their luggage up steep roads to the town perched high above the coastlines" (Markwell, 2015, p. 9; see also Vasilopoulou, Chapter 10). This, Markwell (2015, p. 9) suggests, "no doubt adds to the rustic charm and perceived authenticity of the holiday for some tourists." Such understandings of the dynamics and value of animal labor to destination economies precipitates the potential for discussions about economic necessity, animal exploitation, and stronger animal worker rights, while situating animal workers as stakeholders in the tourism system (see Garcia-Rosell & Tallberg, Chapter 7).

As a component of value-added work, we have also included animals as co-workers to humans. In this way, the labor performed by animals happens in conjunction with humans to improve or add quality to a particular task. The animal is not necessary to achieve the task, but by working together there is improved efficiency, as well as the potential to form strong interspecies relationships. Such value-added work builds upon long histories of human-animal interactions, particularly in cases where animals are domesticated and have been bred for particular traits. As a result, the animal's unique skills beyond any human ability become especially valuable and trusted. While such cases demonstrate animals who have been bred and trained, and as such have reduced agency in the relationship, they also illustrate situations in which humans sometimes must turn over control and put their trust in the animal's judgement (see Hoarau-Heemstra & Nazarova, Chapter 8).

Examples of human-animal co-working are dominated by highly social domesticated animals. Indeed, we can observe an increase in the employment of dogs at security checkpoints, as customs agents, and even other jobs that adapt their herding skills. For example, Piper, a Border Collie, gained notoriety for his work as a wildlife control agent at an airport in Traverse City, Michigan (USA). He worked in partnership with the airport's operations manager to keep the runways clear of birds, which pose a hazard to planes taking off and landing. While this fits firmly within Urbanik's (2012) category of service work, Coulter's (2016) concepts of communication and body work are also essential to Piper and his handler preforming their job, by using his heightened sense of smell or chase and/or herding instincts, all while communicating with his human co-worker. However, in addition to value-added work, the behind the scenes nature of this type of safety/security also demonstrates the overlap with hidden labor. While Piper, specifically, is a case where the behind the scenes, day-to-day, safety work of this dog was actually pushed into the spotlight through the use of social media, in fact, most animals, dogs in particular, who labor in safety and security roles are rarely seen. So while the animal as co-worker adds a dimension to completing tasks that humans simply cannot do on their own, the majority of the public is not aware of their work.

Thus, value-added animal labor fits within Urbanik's (2012) category of service work, but also includes many aspects of Coulter's (2016) work, including care, communication, emotion, body, and dirty work. Importantly, we must also recognize that like human work, the types of labor performed are often diverse. As such, animals who are primarily employed for their value-added work, might be co-opted into performances for entertainment due their public popularity, as such Piper (airport wildlife control officer) who developed a media following and did press engagements. Moreover, such animals might also be engaged in what we term hidden labor, where their labor happens behind the scenes or their image or even livelihoods are rhetorically stretched to broader political agendas far beyond their local agency.

Hidden labor

We live in a visual world in which images are increasingly used as means of communication and social relations (Debord, [1967]2009). This has been particularly observed in the context of tourism (Desmond, 1999; Urry & Larsen, 2010). While the vast majority of animals in the tourism industry perform some form of physical labor, we have also observed a number of cases in which these same animals are also a part of symbolic labor. As described in the "performative work" and "value-added work" approaches above, the symbolic value of animals in the tourism industry means that their labor adds to destination image in various ways. However, we suggest there is another way of seeing the labor that these animals perform. By employing the concept of "hidden labor" we aim to draw attention to its oxymoronic character by highlighting what might not be so obviously described as "labor" *per se*, but we suggest is work that is essential to survival, and yet sometimes also simultaneously linked to demise of a species more broadly. Thus, we draw on aspects of the concepts of "invisible labor" and "hidden work" from gender and race studies (see Crain et al., 2016) and "hidden labor" from critical disability studies (see Scully, 2010). In addition to the safety/security work described above, we observe three distinct instances of hidden animal labor in the tourism industry. However, we speculate that there might be many more that we are overlooking.

The first type of behind the scenes animal labor that is also a form of value-added work is ecosystem services (Dietz & York, 2015). Nature-based, wildlife, and ecotourism all rely on tourists' perceptions of high-quality natural environments. Resilient ecosystems, however, require a complex balance of species, many of which are not seen or remain uninteresting to tourists, visitors, and outdoor recreationalists. Thus, nature-based tourism foregrounds the symbolic value of popular species, but also relies on the value-added and behind the scenes work of all species to perform their ecosystem services. For example, the media popularity of "shark week" has inspired an influx of tourists on Great White Shark tours, which at the same time purport to educate the public about their ecosystem value and claim that profits go to their conservation (Chivell, 2018). As top predators within marine ecosystems, Clua and Pascal (2014) map the "total economic value of sharks" identifying direct-use value (consumptive and non-consumptive), indirect-use value (ecosystem services), and non-use value (natural and cultural heritage). They suggest that while ecosystem services are not easily monetized in our capitalist economies, the performative work that sharks do within the tourism industry also secures their important place within the marine ecosystem while simultaneously maintaining a resilient ecosystem that is crucial to the sustainability of the tourism industry. Ironically, however, the very popularity of some animal ambassadors can have the effect of actually endangering their habitats and well-being, as tourists are willing to trek off the beaten path and into fragile ecosystems to encounter them, not to mention the carbon emissions they create from the journey itself.

Beyond the social discourses and tourists' geographic imaginaries of destinations, engagements with animals through tourism can also be a part of a second type of hidden labor – political work. The changing tides of popular media and environmental conservation move animals in and out of the spotlight as ambassadors of causes, thereby influencing the types of human-animal interactions sought after through tourism. Particular species ebb and flow in popularity, especially when they relate to natural disasters or last chance tourism. Their images are co-opted for local, national, and global conservation campaigns – look no further than the very symbol of the World Wildlife Foundation and most popular visitor attraction in Chengdu, China – pandas. As a result, their very livelihoods, whether as wild or captive animals, have the potential to influence policy at numerous scales.

While political work is, indeed, quite visible, the ways that humans have crafted animal ambassador roles extracts their everydayness in order to advance broader (human) social movements minimizes the animal's agency. In renders their daily livelihoods a performance with implications far beyond their actual habitats. For example, for pandas their continued performance of stereotypical "panda-ness" is essential to the survival of the species, particularly as their wild numbers are so few and as a result rely on breeding programs to maintain the species (Martin-Wintle, et al., 2015). Moreover, as Bueddefeld and Benbow (Chapter 12) demonstrate for polar bears, as a leading ambassador species of global climate change, the greater degree to which polar bears perform "climate disaster" and "species extinction" – in other words, they actually live these conditions – the more value is placed upon them as an object of last chance tourism. As a result, we suggest that animal ambassadors' political work demonstrates an overlap of both hidden labor and performative work.

Finally, a third type of hidden labor is animal-to-animal social, emotion, and care work. Animals have social and emotional needs, and as such rely on human and non-human animals to help fulfil them. Thus, we can think of the ways animals perform care and emotional work (Coulter, 2016) for one another. Within captive environments animals may provide emotional support and share affection with one another, which goes unnoticed or not fully realized by humans. Indeed, Speiran (Chapter 11) observes a number of instances of captive primates caring for another within enclosures, including grooming, feeding, playing, and affection, as well as captive-wild primate interactions between individuals inside and outside of the sanctuaries. Such moments, Speiran notes, were sometimes quite subtle, leading her to ask what behaviors and types of care work were happening beyond human perception. However, some types of animal-to-animal care work might also be central to visitors' expectations of animal behavior, which suggests another way that hidden labor can sometimes overlap with performative work. Indeed, Guo (Chapter 13) draws attention to tourists' expectations of pandas and observes disappointment among tourists who do not witness pandas playing and interacting with one another. As further evidence of the porous and relational nature of these categories of animal work, we can also draw attention to the examples of cross-species animal friendships, where their care and emotion work catches the attention of humans (Dagg, 2011), which then becomes a visitor attraction in itself. As a result, hidden labor can transform into performative work (Urbanik, 2012).

Structure of the book

This book follows the broad categorical approaches of animal labor observed in the tourism industry and outlined above: performative work, value-added work, and hidden labor. While these approaches provide a way to group and structure the book, it is important to note that most chapters also demonstrate the porous boundaries of animal "work." Indeed, just like human workers, animals in the tourism industry often perform many types of work.

To help the reader establish a critical perspective from which to approach the issue of animal labor in tourism, we begin in Part 1 with two theoretically driven chapters. Both ask the reader to consider the implications of what it means to identify animals as "workers." In particular, Dashper takes a more-than-human approach to differentiating work and non-work in the tourism industry. In so doing, she delineates a number of factors that influence non-human workers: species of animal (wild versus domesticated), level of captivity/freedom, performances and expectations, and degree of agency. In the following chapter, Carr further enriches the conversation by adopting an animal-centric perspective while interrogating the human construct of work and its dialectic, leisure. The author confronts head-on the challenges and limitations of our ability to think "animal-centrically," which he argues all the more necessitates a renewed interest in animal welfare and well-being discussions within the industry.

Moving into Part 2, the chapters demonstrate the performative work of animals in the tourism industry, with animals working as key assets of a destination image and tourist motivation and experience. To begin, Venegas and López detail the history of the Tijuana "zonkey," a donkey painted in stripes who has been used as tourist transportation, amusement, and photo opportunity for decades. In the next chapter, the same authors shed light on the macabre tourist attraction and cultural practice of bullfighting in which the bulls are bred and raised to perform their own death for the amusement of spectators. This is followed by Tully and Carr who take a critical investigation of rural tourism, which so commonly associates farm animals with a rural idyll. However, the authors demonstrate the ethical shortcomings of marketing tourist-farm animal interactions that do not attend to the complex physical, mental, emotional, and social interests of the animals as active participants.

Part 3, then, addresses more subtle layers of working relations among humans and animals that contribute to value-added work within the tourism industry. García-Rosell and Tallberg expand upon stakeholder theory by integrating a feminist framework. Through this lens the authors are able to examine the extent to which an ethics of care includes huskies, reindeer, and horses as stakeholders in Lapland's tourism industry. In a similar context, Hoarau-Heemstra and Nazarova focus on Russian sled dogs and Icelandic horses. However, they more specifically examine the instances in which leadership is distributed among guides, tourists, and animals, thereby demonstrating animal agency. Following on from these cases of humans-animals co-working, Hansen, Rickly, Halpern, and McCabe explore the guide dog-human relationship in the tourism context. The authors describe the experiential dimensions of traveling with vision impairment and the extent to which guide dogs add to their experiential benefits and assist in overcoming barriers. Finally, Vasilopoulou explores the local debates regarding the use of donkeys as pack animals in Santorini, Greece. The author uncovers diverse perspectives that relate the use of donkeys to cultural tradition and heritage, as well as valuable economic resource, while also finding local voices who wish to liberate the donkeys of their beast of burden status.

From explorations of various co-working in value-added human-animal contexts, in Part 4 the authors shed light on the complex roles that animals play in both tourism destinations and broader political agendas. While we term this "hidden labor," that does not make it entirely invisible. Rather, the labor examined here is at the intersections of survival, politics, and care. Speiran's ethnographic work in Costa Rican primate sanctuaries articulates the emotion work that primates perform with and for themselves, whereas educational and entertainment work is directed towards the visiting audiences at the sanctuaries, and finally ecological work conceptualizes their roles as ambassadors for their species and conservation more broadly. Similarly, Bueddefeld and Benbow extrapolate the broader social discourses assigned to polar bears as ambassadors of climate change conservation initiatives. In particular, in situ (wild) and ex situ (zoo) polar bears are compared for their perceived responsibilities and value in their ambassadorial roles as lively commodities of the climate change economy in which their very existence is a form of labor despite the decline of the species. Finally, Guo specifically addresses the linguistic interactions of tourists encountering giant pandas, revealing that the "work" of the pandas revolves around being and acting/performing "cute." As such, conservation of the pandas is driven by the success of this work and how well it satisfies the tourists' self-interest.

Finally, Part 5 offers some reflections on working animals of tourism. During a multidisciplnary symposium on animals in tourism held in Uppsalla, Sweden, in 2019, Kline hosted an open discussion on the book's theme with willing participants. The conversation focused entirely on the topic of animal labor in tourism: its relevance and importance, the research priorities within its broad span of meanings, the current state of research, and potential strategies for achieving a research agenda. This chapter is the resulting commentary of that conversation. And lastly, within the afterword, Fennell shines a spotlight on the notion of animal suffering and positions it alongside an ancient story of our justice system. Using the Greek tale of Aeschylus' *Oresteia,* he illuminates the lack of justice for animals – tragically the same state now as when the trilogy was originally written. Let us hope that as more conceptual and empirical work is done around animals in tourism, we will learn how to build an animal-centric justice system.

References

Bekoff, M. & Nystrom, J. (2004). The other side of silence: Rachel Carson's views of animals. *Zygon®*, *39*(4), 861–884.

Birke, L. & Hockenhull, J. (Eds.). (2012). *Crossing boundaries: Investigating human-animal relationships* (Vol. 14). Brill.

Burns, G. L. (2015). Animals as tourism objects: Ethically refocusing relationships between tourists and wildlife. *Animals and tourism: Understanding diverse relationships*, *67*, 44.

Carr, N., & Broom, D. M. (2018). *Tourism and animal welfare*. CABI.

Carr, N., & Young, J. (Eds.). (2018). *Wild animals and leisure: Rights and wellbeing*. Routledge.

Carter, B. & Charles, N. (2013). Animals, agency and resistance. *Journal for the Theory of Social Behaviour*, *43*(3), 322–340.

Caton, K. (2012). Taking the moral turn in tourism studies. *Annals of Tourism Research*, *39*(4), 1906–1928.

Chivell, W. (2018). The tourism industry and shark welfare. *Tourism and Animal Welfare, eds N. Carr and DM Broom (Wallingford: CABI)*, 97–101.

Clua, E., & Pascal, N. (2014). Shark-watching ecotourism in the Pacific islands: A move towards "payments for ecosystem services". *SPC Fisheries Newsletter*, *144*, 30–34.

Cohen, E. (2009). The wild and the humanized: Animals in Thai tourism. *Anatolia*, *20*(1), 100–118.

Cohen, E. (2012). Tiger tourism: from shooting to petting. *Tourism Recreation Research*, *37*(3), 193–204.

Cohen, E., & Fennell, D. (2016). The elimination of Marius, the giraffe: Humanitarian act or callous management decision?. *Tourism Recreation Research*, *41*(2), 168–176.

Coulter, K. (2016). *Animals, work, and the promise of interspecies solidarity*. Palgrave Macmillan.

Cousquer, G., & Allison, P. (2012). Ethical responsibilities towards expedition pack animals: The mountain guide's and expedition leader's ethical responsibilities towards pack animals on expedition. *Annals of Tourism Research*, *39*(4), 1839–1858.

Crain, M., Poster, W., & Cherry, M. (2016). *Invisible labor: Hidden work in the contemporary world*. Univ of California Press.

Dagg, A. I. (2011). *Animal friendships*. Cambridge University Press.

Dashper, K. (2017). Listening to horses: Developing attentive interspecies relationships through sport and leisure. *Society & Animals*, *25*(3), 207–224.

D'cruze, N., Machado, F. C., Matthews, N., Balaskas, M., Carder, G., Richardson, V., & Vieto, R. (2017). A review of wildlife ecotourism in Manaus, Brazil. *Nature Conservation*, *22*, 1–16.

Debord, G. ([1967]2009). *Society of the Spectacle*. Sussex: Soul Bay Press Ltd.

Dietz, T. & York, R. (2015). Animals, capital and sustainability. *Human Ecology Review*, *22*(1), 35–54.

Desmond, J. C. (1999). *Staging Tourism: Bodies on Display from Waikiki to Sea World*. Chicago: The University of Chicago Press.

Fennell, D. A. (2012). Tourism and animal rights. *Tourism Recreation Research*, *37*(2), 157–166.

Fennell, D. A. (2015). Ethics in tourism. In *Education for sustainability in tourism* (pp. 45–57). Springer, Berlin, Heidelberg.

Frost, W. (2011). *Zoos and Tourism: Conservation, Education, Entertainment?* Channel View.

Hewson, C. J. (2003). What is animal welfare? Common definitions and their practical consequences. *The Canadian Veterinary Journal*, *44*(6), 496.

Higgin, M. (2012). Being guided by dogs. In L. Birke & J. Hockenhull (Eds.), *Crossing boundaries: Investigating human-animal relationships* (pp. 73–88): BRILL.

Higginbottom, K. (Ed.). (2004). *Wildlife tourism: Impacts, management and planning*. Common Ground Publishing.

Hill, S. P., & Broom, D. M. (2009). Measuring zoo animal welfare: theory and practice. *Zoo Biology: Published in affiliation with the American Zoo and Aquarium Association*, *28*(6), 531–544.
Hribal, J. C. (2007). Animals, agency, and class: Writing the history of animals from below. *Human Ecology Review*, 101–112.
Jenkins, O. H. (2003). Photography and travel brochures: the circle of representation. *Tourism Geographies*, *5*(3), 305–328.
Kistler, J. M. (2004). Introduction. In Kistler, J.M. (Ed.) *Animals Are the Issues: Library Resources on Animal Issues*. New York. The Haworth Press: 1–6.
Kline, C. (Ed.). (2018a). *Animals, Food, and Tourism*. Routledge.
Kline, C. (Ed.). (2018b). *Tourism experiences and animal consumption: Contested values, morality and ethics*. Routledge.
Knapp, C. R., Hines, K. N., Zachariah, T. T., Perez-Heydrich, C., Iverson, J. B., Buckner, S. D., & Romero, L. M. (2013). Physiological effects of tourism and associated food provisioning in an endangered iguana. *Conservation Physiology*, *1*(1).
Lovelock, B. (2008). An introduction to consumptive wildlife tourism. In *Tourism and the consumption of wildlife. Hunting, shooting and sport fishing* (Vol. 9, p. 290). Routledge Taylor and Francis Group London and New York.
McFarland, S. E. & Hediger, R. (2009). Approaching the agency of other animals: An introduction. In *Animals and Agency* (pp. 1–20). Brill; Leiden, The Netherlands.
Markwell, K. (Ed.). (2015). *Animals and tourism: Understanding diverse relationships* (Vol. 67). Channel View Publications.
Martin-Wintle, M. S., Shepherdson, D., Zhang, G., Zhang, H., Li, D., Zhou, X., & Swaisgood, R. R. (2015). Free mate choice enhances conservation breeding in the endangered giant panda. *Nature communications*, *6*(1), 1–7.
Mann, J., Senigaglia, V., Jacoby, A., & Bejder, L. (2018). A comparison of tourism and feeding wild dolphins at monkey Mia and Bunbury, Australia. *Animal welfare and tourism*. CABI: Oxfordshire.
Nance, S. (2013). *Entertaining elephants: Animal agency and the business of the American circus*. Baltimore, MD; John Hopkins University Press.
Newsome, D., Dowling, R.K., & Moore, S.A. (2005). *Wildlife tourism*. Channel View Publications.
Notzke, C. (2019). Equestrian tourism: animal agency observed. *Current Issues in Tourism*, *22*(8), 948–966.
Orams, M. B. (2002). Feeding wildlife as a tourism attraction: a review of issues and impacts. *Tourism management*, *23*(3), 281–293.
Rodger, K., Moore, S., & Newsome, D. (2009). Wildlife tourism, science and actor network theory. *Annals of Tourism Research*, 36(4): 645–666.
Shackley, M.L. (1996). *Wildlife tourism*. Cengage Learning EMEA.
Shani, A. & Pizam, A. (2008). Towards an ethical framework for animal-based attractions. *International Journal of Contemporary Hospitality Management*, *20*(6), 679–693.
Scully, J.L. (2010). Hidden labor: Disabled/nondisabled encounters, agency, and autonomy. *International Journal of Feminist Approaches to Bioethics*, 3(2), 25–42.
Špinka, M. (2019). Animal agency, animal awareness and animal welfare. Animal Welfare, 28(1), 11–20.
Steward, H. (2009). Animal Agency. Inquiry: an interdisciplinary journal of philosophy, 52 (3). pp. 217–231.
Tully, P. A., & Carr, N. (2020). Presenting the donkey at the seaside: a move towards tackling speciesism in the tourism industry. *Annals of Leisure Research*, 1–18.
Urbanik, J. (2012). *Placing animals: An introduction to the geography of human-animal relations*. Rowman & Littlefield.

Urry, J. & Larsen, J. (2011). *The tourist gaze 3.0*. Sage.
Winter, C. (2020). A review of animal ethics in tourism: Launching the annals of tourism research curated collection on animal ethics in tourism. *Annals of Tourism Research, 84*.
Yerbury, R., Boyd, W., Lloyd, D., & Brooks, A. (2017). Right to leisure? Refocusing on the dolphin. *Annals of Leisure Research, 20*(3), 368–385.
Young, J. & Carr, N. (Eds.). (2018). *Domestic animals, humans, and leisure: Rights, welfare, and wellbeing*. Routledge.

Part 1: **Theoretical considerations**

Katherine Dashper

2 Conceptualizing non-human animals as "workers" within the tourism industry

Introduction

Non-human animals have long been involved in tourism, whether that be as exhibits to be gazed upon in zoos and aquaria, wildlife to be sought out and either watched or hunted in their natural habitats, or as providers of services ranging from transport to food. While the importance of animals to so-called "consumptive" and "non-consumptive" wildlife tourism has been increasingly acknowledged, and the capacity for tourism to both harm and potentially protect non-human life is beginning to be explored (Highman & Shelton, 2011; Margaryan & Wall-Reinus, 2017), there has – with few exceptions (e.g. Dashper, 2020a) – been less attention paid to the labor of animals in the service of tourism organizations and their roles servicing the wishes of paying guests. This is an important omission in understanding both animal tourism and tourism work, as non-human animals perform a variety of important roles in the tourism industry which should be classed as "work." Acknowledging the labor of animals, and their efforts as work, has important implications for how we understand tourism work and the roles of non-human animals in the global tourism industry.

This chapter begins to redress this omission by considering what it might mean to classify non-human animals as "workers" in the tourism industry and to recognize their labor as work. Human labor in the tourism industry is often characterized by poor working conditions, seasonality, low pay, and low skill roles (Poulston, 2009); yet, tourism work also offers many people in precarious positions viable economic opportunities, and women in particular may benefit as they are more likely to be business owners in tourism than in any other industry (UNWTO, 2011). Tourism thus offers opportunities for economic growth and prosperity but is also a sector in which numerous groups are marginalized and rendered vulnerable, through precarious and often exploitative working conditions. If many human workers are marginalized in and through tourism work, what about non-human workers? In an anthropocentric society where human interests take precedence over non-human, what does this mean for the position of non-human workers in organizations where even many humans are often subject to poor conditions and low status? Are non-human animals even more vulnerable as workers in the tourism industry? In this chapter, I begin to consider some of these issues in relation to non-human animal workers in the global tourism industry, and some of the consequences of classifying animals as workers, in terms of (non-human) worker rights and responsibilities.

The chapter begins with a discussion of work and some of the conceptual issues that arise when trying to define and differentiate "work" from "non-work." I then

https://doi.org/10.1515/9783110664058-002

introduce a more-than-human perspective to these debates and consider how moving beyond anthropocentric and speciesist definitions of work and worker involves recognizing and valuing the labor of non-human animals. This can contribute to opening up debates around work and working conditions for all in the tourism industry – human and non-human.

Conceptualizing work

According to Coulter (2016b), work is a crucial political terrain, and so excluding non-human animals from the category of "worker" is a political act that serves to erase and render invisible the labor done by other animals in the service of human organizations, in and beyond tourism. Debates in the sociology of work and associated fields illustrate that the very concept of work itself, and who and what can be defined as a "worker" and "working," are complex, debatable, and historically and culturally specific (Lyon, 2010). Traditional Marxist-inspired approaches tend to associate "work" with paid employment but, as feminist theorists have long pointed out, this is not sufficient to account for the variation in work undertaken by different actors in different contexts. Glucksmann (1995; 2009) argues for a more relational understanding of work that rejects dualisms of work/leisure, public/private, paid/unpaid as analytically narrow and therefore lacking in utility. The dualism between human/non-human animal is another binary that needs to be deconstructed in relation to concepts of work.

Narrow definitions of work related solely to paid employment often fail to recognize broad swaths of work that are integral to the functioning of society. Williams and Nadin (2012) identify three categories of work not covered by a narrow concentration on paid labor: unpaid work done in the home, voluntary and community work, and undeclared work that may be monetized but is informal and not registered with tax services. Failing to broaden our conceptualization of work to include these kinds of activities leads to an undervaluation of some forms of labor and, by association, those who perform those roles. This has long been the case for the underappreciation of the importance of domestic labor and care, which have often been seen as "not real work" in comparison to waged work – i.e., "real work." This is underpinned by the assumption that the public sphere of work is firmly differentiated from the domestic sphere of the home and leads to the creation of two distinct types of worker: those assumed to be able to put the needs of work first and to be able to separate from the domestic sphere (mainly men) and those who cannot (mainly women) (Bourne & Calàs, 2013). Women are thus positioned at a distance from the ideal worker who is presumed to be gender neutral but is instead implicitly gendered male, leaving women marginalized and disadvantaged in relation to attempts to embody organizationally valued norms. Many people are also marginalized from notions of the ideal worker through their "otherness" from the norm in relation to race and ethnicity, socio-economic status and (dis)ability. Narrow definitions of work, and by extension what constitutes an ideal worker, is a contributor to widespread and persistent inequality in almost all sectors of the workforce, including tourism (Costa et al., 2017; Dashper, 2020b).

According to Parry et al. (2005), we need to adopt a broader understanding of work as "being embedded in other domains and entangled in other sorts of social relations" (p.4) if we are to recognize "the way in which inequalities manifested in one domain have a reiterative relationship with behaviors and values in another domain" (p.4–5). If women, people of color, and those disadvantaged by class position, disability or other factors, are marginalized by narrow definitions of work and find themselves struggling to embody norms of the ideal worker, then non-human animals are also likely to be side-lined and excluded from full recognition as productive and valued workers as they are even further from the normative ideal of the adult human male worker. As marginalized human workers have long experienced, this has both material and symbolic consequences in terms of work quality, conditions, and status.

Non-human animals as workers

Non-human animals have been conspicuously absent from studies of work, organization, and management (Labatut, Munro, & Desmond, 2016; Sayers, Hamilton, & Sang, 2019). Only very recently has this begun to be addressed, and consequently the labor and working experiences of non-human animals are undertheorized. Several recent journal special issues have begun to address this gap, although focus remains predominantly on the experiences of working with animal death and suffering, such as in the slaughter industry (McCabe & Hamilton, 2015; McLoughlin, 2019), veterinary practice (Clarke & Knights, 2019; Sanders, 1994), and animal cruelty investigation (Coulter & Fitzgerald, 2019), or caring work performed with, for, and by animals (Charles & Wolkowitz, 2019; Coulter, 2016a; Taylor, 2010). This, and other research, is beginning to unpack some of the issues associated with non-human animals as workers in their own societies, as well as in human organizations, and the complexities of human-non-human relationships and interactions in the context of work and organizations (Davies & Riach, 2019).

This emergent scholarship highlights how human and non-human work are often mutually constitutive, helping shape working experiences across species divides (Knight & Sang, 2019). Importantly, human workers who labor with and for animals are often subject to low status and poor working conditions, and the marginalization of human and non-human worker often go together. Human workers involved in agriculture and hands-on work with animals appear to suffer a form of contamination by these associations, linked through animal-related "dirty work" to muck and mess and consequently accorded low status (Coulter, 2017; Mitchell & Hamilton, 2018; Sanders, 2010). Porcher (2011) and Coulter (2016b) both argue that human and non-human suffering are often linked in the workplace, and that many contemporary organizational practices alienate both human and non-human workers, leading to physical, mental, and moral suffering. As such, capitalist work practices may be causing serious harm across species boundaries.

Work with, for, and by animals is also often feminized, contributing further to marginalization and low status in the masculine world of work and organizations (Coulter & Fitzgerald, 2019). Women and femininity have long been associated with animals and animality, in pejorative ways that delegitimize both (Adams, 2018; Birke, 2007; Dashper, et al., 2018). Emergent research on non-human work and organizations illustrates that these associations remain strong and serve to further marginalize both women and non-human workers. Focusing on the labor and experiences of non-human animals may thus highlight many of the injustices experienced by a variety of human workers as well (Ridout, 2004).

While much focus has been placed on important issues to do with marginalization, discrimination, and abuse, this research is also beginning to illustrate the agency of non-human animals to actively shape organizational practices and experiences. Although animals are often made invisible in contemporary organizations, even when they are central to their practices, such as in the food production industry (Evans & Miele, 2012), they are active living beings and often make their presence known in ways that may surprise, delight, and sometimes frustrate human organizational members and researchers. Similarly, research on relationships between people and the dogs whose labor acts as a form of assistance (whether in terms of vision, emotional support, or other roles) illustrate the transformative effect dogs can have on the lives of the humans with whom them work (Rickly, 2018; Sanders, 1999, 2000). Sage et al. (2016) show that human and non-human organizing are often entangled. In their study of construction, they illustrate how the human-focused practice of the managed production of the built environment was often frustrated by the physical presence and actions of animals within those spaces. O'Doherty's (2016) fascinating ethnography of airport workers illustrates how an originally uninvited organizational member – Olly the cat – actively shaped (human) organizational culture in ways surprising and unexpected to both management and the researcher.

Although it is important to acknowledge that non-humans have some agency, the status of non-human animals in organizations is highly ambivalent. Even animals with distinct roles acknowledged to be important to human goals, such as police dogs, occupy a liminal position between objects deployed to protect, assist, and serve humans and individual subjects capable of developing bonds with others and with distinct needs and desires (Sanders, 2006). Knight and Sang (2019) discuss this ambivalence in the context of police dogs' non-working lives. Whilst active in police service many dogs live at home with their human police handler, who often report close relational bonds between the two. However, on "retirement" from active police duty many dogs are deemed unsuitable for "civilian life" and euthanized despite being physically well and healthy. Dogs are valued as co-workers when they are useful to humans, but easily discarded when they are no longer deemed capable of providing a function valued in purely anthropocentric terms. The experiences of horses show similar ambivalence. Dashper (2014) discusses how elite competition horses are valued for their individual personalities, abilities, and dispositions, yet also treated as commodities to be bought and sold according to human economic desires. Non-human animals may thus be valued for their working capacities and usefulness to humans, yet also marginalized and granted lower status by virtue of their non-humanness which is assigned less worth in the anthropocentric world of work. More-than-human perspectives, which have been adopted in many of the studies cited in this section, expose this anthropocentrism and look for ways to begin to overcome some of the shortcomings of valuing non-human work and workers only in relation to their usefulness to human interests.

Introducing more-than-human perspectives

Most research on animals and tourism takes an anthropocentric focus, concentrating on human-defined goals about maximizing tourist satisfaction or managing wildlife-tourist interactions to minimize conflict, but ensure tourism practices can continue largely unfettered (e.g. Ballantyne, Packer, & Sutherland, 2011; Skibbins et al., 2012). Even research focusing on animal ethics and welfare tends to take a broadly anthropocentric view, focusing on instrumental arguments about balancing welfare and conservation with the needs of the tourism industry and rarely questioning if tourism itself may need to change in favor of addressing the needs of other species (Cunningham, Huijbens, & Wearing, 2012; Larm et al., 2017). A few studies are beginning to recognize non-human animals as actors in tourism encounters, with their own distinct rights, needs, and behaviors defined separately from their relationship with human needs and practices, and thus questioning some of the workings of the tourism industry in relation to other animals (e.g. Cohen & Fennell, 2016; Yudina & Grimwood, 2016). However, the anthropocentric focus of much wildlife and other animal-related tourism research is a factor in the lack of attention paid to date to the labor and working lives of non-human animals within the tourism industry. The work of non-human animals has so far gone largely unexamined, taken for granted, and undervalued as research focuses on human-centric issues and priorities.

One strand of research in tourism studies that does challenge anthropocentrism draws on new materialism and predominantly Actor Network Theory (ANT) to consider how human and non-human, animate and inanimate actors produce and maintain tourism practices. ANT approaches argue that "the ability to act or create an effect does not stem from being human, but is accomplished by enacting and being entangled in a web of connections" (van der Duim, Ren, & Jóhannesson, 2017, p. 141). This resonates with the more relational view of work advocated by Glucksmann (1995, 2009) outlined above and leaves analytical space to consider a broad range of working practices and experiences, including those of non-human animals. However, as Cohen and Cohen (2017) argue, ANT and material studies "have a tendency to dichotomize human and non-human actors, leaving the status of animals unclear, as they become lumped in with material objects" (p. 9). Van der Duim, Ampumuza, and Ahebwa's (2014) study of gorilla tourism in Uganda illustrates this point. They use ANT to trace different actor networks between the gorillas, forest, humans, and other objects but in so doing the gorillas themselves "seem to lose their status as sentient beings," positioned as actants that influence events through their interactions with other actants but losing any sense of them as embodied creatures with individual selves, lives, and needs (Cohen & Cohen, 2017, p.9). So, while ANT offers an interesting approach to studying tourism, and does important work in decentering human priorities and interests, it may not be the most effective tool to examine the working lives and experiences of non-human animals which may still disappear from focus in ANT-inspired research.

Instead, I suggest the need to turn to the field of human-animal studies to find the tools necessary to not only decenter human priorities but also to try and acknowledge and account for the experiences of non-human animals working in the service of the tourism industry. Such an approach recognizes that "societies are broader than the human" (Peggs, 2013, p. 591) and we need to consider how human and non-human lives and experiences entangle and co-constitute each other. Haraway's (2003) insights on the importance of considering the "co-habitation, co-evolution and embodied cross-species sociality" (p. 4) of our lives with companion species is informative here as it draws attention to the ways in which human and non-human life cannot be considered as separate entities, but as deeply entwined mutually constitutive experiences.

A more-than-human perspective advocates the importance of trying to see issues and experiences at least partly on the terms of non-human, as well as human, animals. This is difficult to achieve in practice, as we are inevitably bound by our own human perceptions and the anthropocentric structures of academic research. However, while our attempts to consider the world from non-human perspectives may be limited, they are not impossible and trying to consider more-than-human perspectives on a given issue or research question can radically change our conclusions. As Wels (2015) so strikingly demonstrates in his re-evaluation of his own ethnographic fieldwork in Zimbabwe, reframing research questions and analytical apparatus to try and consider issues with the interests of non-human animals in mind can force us to question our practices and their implications for the non-human world. This is important if we are to try and rebalance our world view in ways that are at least slightly less anthropocentric and to acknowledge that non-human species have interests and rights that are as important as our human ones.

Non-human animals as workers in tourism

Not every non-human animal involved in the tourism industry can be classed as a worker, even when a broad and relational view of the concept of work is adopted, as not all animals are engaged in labor that helps sustain the workings of tourism organizations. For example, I would argue lions living on the plains of the Serengeti are not workers in the tourism industry, although they do undoubtedly labor in service of their own pride and conspecifics. Although these lions have a loose relationship with the tourism industry, as they provide a key sight for many tourists and tour companies to seek out and consume visually, they are not in an active relationship with other tourism actors focused on the act of *deliberately* (re)producing an aspect of the tourism product. They receive few direct benefits or outcomes from interactions with people and tourism, and have limited restrictions and expectations placed on them in return. In contrast, lions in a zoo could be considered workers as they are necessarily bound by the requirements of this particular tourism organization and their behaviors and actions dictated by the demands and constraints of the organization. In return, lions in the zoo receive food, shelter, and hopefully appropriate veterinary and other forms of care, but find their behavior curtailed and managed by the zoo staff and regime. Lions in so-called breeding centers that are hand-reared and used for photo opportunities with tourists are also workers, albeit in a more active way than the lions in a zoo. The breeding center lion is likely to have been bred and reared in captivity for the specific purpose of interacting with tourists for photo opportunities, lion walks, and ultimately hunting (Wharton, 2019). These lions will also receive food and shelter in return for their efforts, and some modicum of care, but the appropriateness of this to their needs is questionable. These lions are subject to strict "feeling rules" that restrict their behavior and rigid expectations of how they should interact with tourists (see Dashper, 2020a). Their labor contributes to the commercial success of the breeding center as a tourism organization, even though they have very little control about their involvement in these activities. Lions involved in so-called "canned hunting" can be considered workers in the same ways as lions at breeding centers, as they too are bred and reared to service human tourists' wishes, and positioned in an unequal relationship where their behavior is heavily curtailed in return for limited provision of food and sometimes protection, at least until their "job" requires them to become prey for tourists paying high sums of money. This brief example of lions in different contexts illustrates that non-human animals are positioned in a variety of roles in relation to the tourism industry, some of which can be seen as work, and some not, and with varying levels of expectation and restriction placed on them. There are also wide differences in relation to what the animals receive back in return for their role in supporting the tourism organization.

The variety of roles performed by non-human animal workers can be further distinguished by variation in other factors that affect the experiences of the animals in those positions. These include:

- *Species of animal*: Is the animal "wild" or domesticated? How habituated to the human world is the animal, and how will this affect his or her ability to adapt to and cope with the human-centric demands of tourism?
- *Level of captivity/freedom*: How confined is the animal, both physically (in terms of space and housing etc.) and behaviorally (in terms of the extent to which the natural behaviors of the animal are curtailed and limited)?
- *Performances and expectations*: What is the animal expected to do, as part of the tourism encounter? Is he or she expected to interact directly with tourists and/or tourism workers? Is he or she expected to perform some kind of action or activity?
- *Degree of agency*: What degree of agency does the animal have to act and take action in the context of the tourism organization? What kinds of management practices are enacted to limit that agency or to shape it in ways to suit the anthropocentric demands of the tourism industry?

Each of these themes is complex and will vary widely across non-human animal and tourism organization, but each is integral to trying to understand the experiences of non-human workers and to appreciate some of the practical and moral aspects of enlisting their labor in the service of the tourism industry. Each of these issues requires careful empirical examination, but here I briefly consider some key aspects related to each point that will affect non-human animal workers in varying ways.

Species of animal

In their political consideration of animal rights, Donaldson and Kymlicka (2011) distinguish between humans and three types of animal: domesticated, wild, and what they call "liminal" – an in-between position, like feral cats and dogs. They liken each type to a minority group that deserves protection and argue that each group has varying levels of rights and responsibilities depending on their relationships to human societies. Domesticated animals have the closest relationship with humans, and have the highest level of rights (in terms of protection from harm, access to food etc.) and also the highest level of responsibility – they must learn (through training) how to interact with humans in ways that do not harm the human. In contrast, wild animal communities should be acknowledged as sovereign and have their boundaries respected and not violated. In return, they should also respect human community boundaries, but can expect no further protection from humans. Finally, liminal animals are located somewhere in between, being granted rights to interact occasionally with humans, free from fear of harm, but cannot expect further protection from humans. In this view of what Donaldson and Kymlicka (2011) term the "zoopolis" human and non-human animals would live in relative harmony, with varying degrees of interaction and co-dependency. In terms of tourism only domesticated animals might be expected to play an active role as a worker, as only they have entered into a form of relationship that entails mutual rights and responsibilities. In contrast, wild animals would be left beyond human intervention, with their territories and habitats respected as sovereign and independent. Clearly, this is not what happens in practice at the moment, but it points to the level of suitability, or not, that different categories of animal have for engaging in work in tourism organizations. Domesticated animals, like dogs and horses, have a close evolutionary and social relationship with humans and so might be expected to adapt more easily to work in tourism organizations, on the proviso that they be granted appropriate rights and protections. Wild animals who are not entered into a co-dependent relationship with humans are likely to suffer most from captivity and involvement in tourism which in no way respects their sovereignty and autonomy. Involving wild animals in tourism, whether that be in zoos, parks, or even in wild settings, is an encroachment on their sovereign status, and so is likely to cause more distress and discomfort to a wild animal than a domesticated one.

Level of captivity versus freedom

This issue is inevitably linked to the one above, and it might be expected that wild animal workers will be more heavily restricted than domesticated ones, due to their different level of habituation to human interaction and associated risks of human-animal interactions. The less "safe" (or more "dangerous") the non-human animal, in anthropocentric terms, the greater the level of captivity and restriction might be expected. However, this will be heavily dependent on the type of tourism encounter sought between tourist and non-human animal worker. Wild animals in zoos, as discussed above in relation to lions, might be subject to heavy restrictions on their movement and ability to express normal behaviors, as in order to keep the zoo guests safe from potentially dangerous-to-humans wild animals severe restrictions on the animals' freedom and mobility are likely, in the form of fences, water, and electric barriers. Moorhouse, D'Cruze, and Macdonald's (2017) research shows that tourists give little thought to animal welfare at wildlife tourism attractions and have very limited knowledge of what "good welfare" might be for different types of captive species. If there is little pressure from tourists to demand high welfare standards for working animals, which would include their physical living conditions and ability to express “normative” species behavior, then it is unlikely that many tourism organizations will foreground this as a key issue for investment when the high costs required may not be matched by increases in tourist satisfaction. The commercial imperatives of capitalist tourism organizations may not drive up welfare standards. This is likely to profoundly affect the experiences of non-human animals living and working in these conditions and exposes their vulnerability and ambivalent status in the tourism industry, where their well-being may often be valued lower than monetary profit.

Performances and expectations

There is wide variation in terms of the level of interaction with tourists, and the types of activities non-human animal workers are expected to engage in, and this will certainly affect their working experiences. Again, this is linked to the point above about the type and status of species of non-human worker. Domesticated animals are often expected to work in tourism in ways similar to how they work in other fields of human society such as sport and leisure. Sled dogs and riding horses can be considered tourism workers, involved in labor that services the tourism organization and provides a product to the paying tourist, and often are required to engage directly with tourists (Bertella, 2014; Danby, 2018; Garcia-Rosell & Tallberg, Chapter 7; Hoarau-Heemstra & Nadezda Nazarova, Chapter 8). These interactions, and associated expected behavior, place considerable demands on domesticated animals (Dashper, 2020a). However, domesticated animals are habituated to interactions with people, and so these demands, while sometimes stressful, may not be completely outside of their realms of experience.

In contrast, wild animals are not habituated to close human interaction, even when they have been born and raised in captivity, and so direct human interaction will be more challenging. Cong et al. (2014) explain how tourists value the experience of actually holding a panda during (often expensive) giant panda tourism encounters in Chengdu, China. Being held by a human is not a routine experience for a giant panda, and so is likely to cause stress amongst many. Swim-with-dolphin experiences offer another site in which wild non-human animals are expected to interact directly with humans, sometimes having to perform actions such as offering a “kiss” or “shaking fins.” These encounters are often stressful and difficult for the dolphins, particularly when they take place in restricted spaces of dolphin captivity (see point above) and when the dolphins cannot choose if or how to interact with people (see Warkentin, 2011). The nature of human-non-human encounters, and the extent to which activities and behaviors expected of non-human animals are outside their normal realm of behaviors, has important implications for the experiences of animal workers in different tourism contexts.

Degree of agency

Questions of non-human animal agency are complex and fraught (see Dashper, 2017). There is no (human) consensus on how we should understand non-human animal agency, but for the purposes of the argument developed here I understand agency as a continuum along which all animals – human and non-human – are positioned (Pearson, 2013). All animals have agency, although this varies in terms of degree and how it may be expressed in different contexts. However, as we all exist in a human-dominated world, the agency of non-human animals is curtailed by their position in this environment. Carter and Charles (2013) argue that agency is relational, and so non-human animals do have agency to act and bring about change in some circumstances, but this is often conditional on human responses as humans are in positions of power relative to non-humans. This is certainly the case for non-human animal workers in the tourism industry.

Domesticated animals, such as horses, are routinely deployed as workers in the tourism industry, and are expected to perform tasks that require close interaction with humans – both tourism staff and tourists. Trail riding horses often have to carry tourists with little or no experience of the complex interspecies act of communication that is riding, and so may act in ways that are confusing, distressing and even painful for the horse. To try and reduce these negative outcomes, the horse does have some agency to act and try to change the situation. For example, the horse may plod slowly to reduce the movement of the unbalanced tourist on his or her back, thus minimizing friction and making the weight easier to carry. Or the horse may snatch the reins from the hand of a tourist who is grabbing and pulling on the horse's mouth, again minimizing pain or discomfort. Here, the horse has some agency to act and bring about a change in circumstances that might make the encounter with the tourist less uncomfortable and stressful for the horse. However, in the human-defined context of tourism the horse's agency is curtailed by the responses of humans. The tourist may respond by pulling back at the reins (and consequently the horse's mouth) in an attempt to regain control or kicking at the horse's sides to encourage him or her to speed up. The tour guide may also intervene and shout at the horse, and/or use a stick to enforce compliance with the wishes of the tourist and expectations of the paid-for tourism encounter. In such ways, horses – and all non-human animal workers in tourism – are constrained by the human-defined requirements of the organization and have their agency curtailed by the more powerful human actors in the tourism encounter (see also Dashper, 2020a).

Conceptualizing non-human animals as workers within the tourism industry

The concept of work needs to be extended to encompass the labor performed by non-human animals. In the tourism industry, a wide variety of animals perform diverse acts in the service of tourism organizations, and so constitute an important part of the tourism workforce. Recognizing non-human animals as workers is an important step towards valuing their labor, but also exposes the vulnerability of animal workers in the tourism industry. In this chapter I have proposed four themes around which non-human animal labor can be conceptualized, and there is now need for empirical research to explore animal work and workers in a variety of contexts.

Recognizing non-human animals as workers in tourism has important implications. As Coulter (2016b) argues, work is a crucial political terrain and so acknowledging animals as workers has implications for how they should be treated and interacted with in the context of tourism practices and organizations. Workers have rights, although these vary widely in different sectors, local and national contexts, and in relation to different types and category of worker (Flanagan, 2006). Contingent workers are in low-status and vulnerable positions, and tourism is a sector with many humans involved on such a basis and existing in precarious situations (Robinson et al., 2019). In the anthropocentric worlds of work and tourism the status of non-human workers is extremely low – they represent the most precarious of all workers due to their ambivalent position in a human-centric world where they are positioned as both subject and object, vulnerable to human whim and expectations, and this shapes their working lives and experiences. Examining the working lives of animals in tourism, and the social, economic, and organizational mechanisms that shape their experiences, is important for the welfare and well-being of animal workers. Through better understanding of their experiences we may be able to ensure that working practices and expectations are reshaped with animal interests in mind. Working towards better welfare may require rebalancing the anthropocentric demands of global tourism and the needs and interest of non-human workers. A more biocentric position would involve considering the interests of all animals – human and non-human – in the design of more humane work that can be beneficial for all (Coulter, 2017). As Nibert (2003, p. 21) argues, “the exploitation of humans and other animals is inextricably tangled” and this is evident in the discussion of work and tourism presented in this chapter. Acknowledging non-human animals as workers in the tourism industry, and critically examining their labor and working lives, may lead to better understanding of work across the sector, contributing to envisioning improved working conditions for all workers, non-human and human.

References

Adams, C. J. (2018). *Neither man nor beast: Feminism and the defense of animals*. London: Bloomsbury Publishing.

Ballantyne, R., Packer, J. & Sutherland, L.A. (2011). Visitors' memories of wildlife tourism: Implications for the design of powerful interpretive experiences. *Tourism Management*, *32*(4), 770–779.

Bertella, G. (2014). The co-creation of animal-based tourism experience. *Tourism Recreation Research*, *39*(1), 115–125.

Birke, L. (2007). Relating animals: Feminism and our connections with nonhumans. *Humanity & Society*, *31*(4), 305–318.

Bourne, K. A. & Calás, M. B. (2013). Becoming 'real' entrepreneurs: Women and the gendered normalization of 'work'. *Gender, Work and Organization*, *20*(4), 425–438.

Carter, B. & Charles, N. (2013). Animals, agency and resistance. *Journal for the Theory of Social Behaviour*, *43*(3), 322–340.

Charles, N. & Wolkowitz, C. (2019). Bringing dogs onto campus: Inclusions and exclusions of animal bodies in organizations. *Gender, Work and Organization*, *26*(3), 303–321.

Clarke, C., & Knights, D. (2019). Who's a good boy then? Anthropocentric masculinities in veterinary practice. *Gender, Work and Organization*, *26*(3), 267–287.

Cohen, E. & Fennell, D. (2016). The elimination of Marius, the giraffe: Humanitarian act or callous management decision? *Tourism Recreation Research*, *41*(2), 168–176.

Cohen, S.A. & Cohen, E. (2017). New directions in the sociology of tourism. *Current Issues in Tourism*, 1–20.

Cong, L., Wu, B., Morrison, A. M., Shu, H., & Wang, M. (2014). Analysis of wildlife tourism experiences with endangered species: An exploratory study of encounters with giant pandas in Chengdu, China. *Tourism Management*, *40*, 300–310.

Costa, C., Bakas, F. E., Breda, Z., Durao, M., Carvalho, I., & Caçador, S. (2017). Gender, flexibility and the 'ideal tourism worker'. *Annals of Tourism Research*, *64*, 64–75.

Coulter, K. (2016a). Beyond human to humane: A multispecies analysis of care work, its repression and its potential. *Studies in Social Justice*, *10*(2), 199–219.

Coulter, K. (2016b). *Animals, work and the promise of interspecies solidarity*. New York: Palgrave Macmillan.

Coulter, K. (2017). Humane jobs: A political economic vision for interspecies solidarity and human–animal wellbeing. *Politics and Animals*, *3*, 31–41.

Coulter, K. & Fitzgerald, A. (2019). The compounding feminization of animal cruelty investigation work and its multispecies implications. *Gender, Work and Organization*, *26*, 288–302.

Cunningham, P.A., Huijbens, E.H. & Wearing, S.L. (2012). From whaling to whale watching: Examining sustainability and cultural rhetoric. *Journal of Sustainable Tourism*, *20*(1), 143–161.

Danby, P. (2018). Post-humanistic insight into human-equine interactions and wellbeing within leisure and tourism. In Young, J. & Carr, N. (eds) *Domestic animals, humans, and leisure*. Abingdon: Routledge, 146–164.

Dashper, K. (2014). Tools of the trade or part of the family? Horses in competitive equestrian sport. *Society & Animals*, *22*(4), 352–371.

Dashper, K. (2017). *Human-animal relationships in equestrian sport and leisure*. Abingdon: Routledge.

Dashper, K. (2020a). More-than-human emotions: Multispecies emotional labour in the tourism industry. *Gender, Work & Organization*, *27*(1), 24–40.

Dashper, K. (2020b). Mentoring for gender equality: Supporting female leaders in the hospitality industry. *International Journal of Hospitality Management*, *88*, 1–8.

Dashper, K., Fenner, K., Hyde, M., Probyn-Rapsey, F., Caspar, G., Henshall, C., & McGreevy, P. (2018). The anthropomorphic application of gender stereotypes to horses. *Anthrozoös, 31*(6), 673–684.

Davies, O., & Riach, K. (2019). From manstream measuring to multispecies sustainability? A gendered reading of bee-ing sustainable. *Gender, Work & Organization, 26*(3), 246–266.

Donaldson, S. and Kymlicka, W. (2011). *Zoopolis: A political theory of animal rights*. Oxford: Oxford University Press.

Evans, A.B. & Miele, M. (2012). Between food and flesh: How animals are made to matter (and not matter) within food consumption practices. *Environment and Planning D: Society and Space, 30*(2), 298–314.

Flanagan, R. J. (2006). *Globalization and labor conditions: working conditions and worker rights in a global economy*. Oxford University Press.

Glucksmann, M. A. (1995). Why 'work'? Gender and the 'total social organization of labour'. *Gender, Work and Organization, 2*(2), 63–75.

Glucksmann, M. A. (2009). Formations, connections and divisions of labour. *Sociology, 43*(5), 878–895.

McCabe, D., & Hamilton, L. (2015). The kill programme: an ethnographic study of 'dirty work' in a slaughterhouse. *New Technology, Work and Employment, 30*(2), 95–108.

Haraway, D. (2003). *The companion species manifesto: Dogs, people and significant otherness*. Chicago, Il.: Prickly Paradigm Press.

Higham, J. E. S. & Shelton, E. J. (2011). Tourism and wildlife habituation: reduced population fitness or cessation of impact? *Tourism Management, 32*(6), 1290–1298.

Knight, C., & Sang, K. (2019). 'At home, he's a pet, at work he's a colleague and my right arm': police dogs and the emerging posthumanist agenda. *Culture and Organization*, 1–17.

Labatut, J., Munro, I. & Desmond, J. (2016). Animals and organizations. *Organization, 23*(3), 315–329.

Larm, M., Elmhagen, B., Granquist, S.M., Brundin, E. & Angerbjörn, A. (2017). The role of wildlife tourism in conservation of endangered species: Implications of safari tourism in conservation of the Arctic fox in Sweden. *Human Dimensions of Wildlife, 23*(3), 257–272.

Lyon, D. (2010). Intersections and boundaries of work and non-work. *European Societies*, 12(2), 163–185.

Margaryan, L., & Wall-Reinius, S. (2017). Commercializing the unpredictable: Perspectives from wildlife watching tourism entrepreneurs in Sweden. *Human Dimensions of Wildlife, 22*(5), 406–421.

McLoughlin, E. (2019). Knowing cows: Transformative mobilizations of human and non-human bodies in an emotionography of the slaughterhouse. *Gender, Work and Organization, 26*(3), 322–342.

Mitchell, L. & Hamilton, L. (2018). Hefted: reconfiguring work, value and mobility in the UK Lake district. *Culture and Organization, 24*(4), 303–317.

Moorhouse, T., D'Cruze, N. C., & Macdonald, D. W. (2017). Unethical use of wildlife in tourism: what's the problem, who is responsible, and what can be done? *Journal of Sustainable Tourism, 25*(4), 505–516.

Nibert, D. (2003). Humans and other animals: Sociology's moral and intellectual challenge. *International journal of sociology and social policy, 23*(3), 4–25.

O'Doherty, D.P. (2016). Feline politics in organization: The nine lives of Olly the cat. *Organization, 23*(3), 407–433.

Parry, J., Taylor, R., Pettinger, L. & Glucksmann, M. (2005). Confronting the challenges of work today: New horizons and perspectives. In L. Pettinger, J. Parry, R. Taylor & M. Glucksmann (eds) *A new sociology of work?* Oxford: Blackwell Publishing, pp.3–38.

Pearson, C. (2013). Dogs, history, and agency. *History and Theory, 52*(4), 128–145.
Peggs, K. (2013). The 'animal-advocacy agenda': Exploring sociology for non-human animals. *The Sociological review, 61*(3), 591–606.
Porcher, J. (2011). The relationship between workers and animals in the pork industry: A shared suffering. *Journal of Agricultural and Environmental Ethics, 24*(1), 3–17.
Poulston, J. M. (2009). Working conditions in hospitality: Employees' views of the dissatisfactory hygiene factors. *Journal of Quality Assurance in Hospitality & Tourism, 10*(1), 23–43.
Rickly, J. M. (2018). Considering service animals in tourism. *Annals of Tourism Research, 71*, 57–58.
Ridout, N. (2004). Animal labour in the theatrical economy. *Theatre Research International, 29*(1), 57–65.
Robinson, R. N., Martins, A., Solnet, D., & Baum, T. (2019). Sustaining precarity: critically examining tourism and employment. *Journal of Sustainable Tourism, 27*(7), 1008–1025.
Sage, D., Justesen, L., Dainty, A., Tryggestad, K. & Mouritsen, J. (2016). Organizing space and time through relational human-animal boundary work: Exclusion, invitation and disturbance. *Organization, 23*(3), 434–450.
Sanders, C.R. (1994). Annoying owners: Routine interactions with problematic clients in a general veterinary practice. *Qualitative Sociology, 17*(2), 159–170.
Sanders, C. (1999). *Understanding dogs: Living and working with canine companions.* Philadelphia: Temple University Press.
Sanders, C. R. (2000). The impact of guide dogs on the identity of people with visual impairments. *Anthrozoös, 13*(3), 131–139.
Sanders, C.R. (2006). 'The dog you deserve': Ambivalence in the K-9 officer/patrol dog relationship. *Journal of Contemporary Ethnography, 35*(2), 148–172.
Sanders, C.R. (2010). Working out back: The veterinary technician and 'dirty work'. *Journal of Contemporary Ethnography, 39*(3), 243–272.
Sayers, J., Hamilton, L. & Sang, K. (2019). Organizing animals: Species, gender and power at work. *Gender, Work and Organization, 26*, 239–245.
Skibins, J.C., Hallo, J.C., Sharp, J.L. & Manning, R.E. (2012). Quantifying the role of viewing the Denali 'Big 5' in visitor satisfaction and awareness: Conservation implication for flagship recognition and resource management. *Human Dimensions of Wildlife, 17*(2), 112–128.
Taylor, N. (2010). Animal shelter emotion management: A case of in situ hegemonic resistance? *Sociology, 44*(1), 85–101.
UNWTO (2011). *Global Report on Women in Tourism 2010: Preliminary findings*. Madrid: World Tourism Organisation.
Van der Duim, R., Ampumuza, C. & Ahebwa, W.M. (2014). Gorilla tourism in Bwindi Impenetrable National Park, Uganda: An actor-network perspective. *Society and Natural Resources, 27*(6), 588–601.
Van der Duim, R., Ren, C. & Jóhannesson, G.T. (2017). ANT: A decade of interfering with tourism. *Annals of Tourism Research, 64*, 139–149.
Warkentin, T. (2011). Interspecies etiquette in place: Ethical affordances in swim-with-dolphins programs. *Ethics & the Environment, 16*(1), 99–122.
Wels, H. (2015). 'Animals like us': Revisiting organizational ethnography and research. *Journal of Organizational Ethnography, 4*(3), 242–259.
Wharton, J. (2019). Cute lion photos are fuelling trophy hunting trade. *The Metro* [online] 19 August. Available from www.metro.co.uk/2019/08/19/cute-lion-photos-are-fueling-trophy-hunting-trade-10591099/ [Accessed 03/12/19].
Williams, C. C. & Nadin, S. (2012). Work beyond employment: representations of informal economic activities. *Work, employment and society, 26*(2), 1–10.
Yudina, O. & Grimwood, B.S.R. (2016). Situating the wildlife spectacle: Ecofeminism, representation and polar bear tourism. *Journal of Sustainable Tourism, 24*(5), 715–734.

Neil Carr

3 Redefining the *work* of working animals in the tourism industry: An animal-centric reflection

Introduction

Animals are defined by humans (Hart, 2003; Young & Carr, 2018). At a basic level, animals are only named because humans have named them. For example, a bear does not see itself as a bear. Rather, humans have defined it as such. Similarly, the various dogs that have shared my life clearly do not see themselves as a human and recognize a synergy between themselves and other dogs. However, they do not see themselves as "dogs." They cannot, because "dog" is a human construct, one created without humans ever explicitly asking dogs for any input. Moving beyond this, the behavioral definitions ascribed to different species are the product of human analysis and resultant boundary making. In this way humans define what is normal and what is abnormal behavior in each species. Indeed, such "norms" derive directly from human observations and interpretations. Furthermore, humans increasingly only allow animals to behave in ways humans have defined as normal. Abnormal behavior is frowned upon unless it is showing learning or similar developmental traits. Such traits are praised in the same way humans praise their children for developing a new productive trait. This praise is, of course, human-defined and focuses on humans' obsession with development and learning, with the constant process of productive becoming.

It is important to recognize that behavioral definitions of animals are not only based on scientific analysis. Instead, much of how humans define animals is based on human biases, imaginaries, fears, and wishes. Through these definitions, humans have sought to define themselves as separate from, and above, *mere* animals. Humans have also sought in defining animals to categorize them. In the process we have identified those that we like, those we hate, and those we choose to ignore (Carr & Broom, 2018). Within the context of this chapter, humans have also defined animals in relation to their ability to work for us: how useful they are to us.

When we think of animals as working, we impose more human constructs on them, such as work, non-work, employment, and employee. At the same time, we link them to the concepts of slavery and the slave. If we follow this line of reasoning, we also impose the idea of leisure on animals, recognizing that leisure and work have long been established as related opposites (Clarke & Critcher, 1985; Page & Connell, 2010). From this, we must recognize the all-encompassing nature of the human-dominated world in which animals live.

https://doi.org/10.1515/9783110664058-003

It is against this background that the chapter seeks to critically deconstruct the concept of the working animal. It begins with a critical examination of the ideas of work, non-work, and leisure. Following on from this, the chapter interrogates the idea of "working animal." In this, we see humans thinking in terms of the *value*, or lack thereof, of animals to them. Based on the foregoing thoughts, the chapter then engages with the question of the welfare of working animals. This is, once again, set alongside ideas of worker protection, which is another human construct.

The working animals discussed in this chapter are mainly, though not exclusively, situated within the human tourism experience. The reason for looking inside and outside of this experience is that the concepts discussed cut across the artificial divide that humans have put up around the holiday experience and the rest of their lives. However, the tourism context is important as it provides an additional and unique perspective to the idea of work as the holiday experience has traditionally been constructed in opposition to ideas of work, at least for the tourists. In this way there is a similarity between the working animal in the holiday environment and the humans working in the tourism and hospitality industries. As consumers of the products of these industries, we *know* that there is a vast array of human employees that facilitate the tourism experience. However, how many tourists give conscious thought to the fact that their leisure is facilitated by the work of others. If this is true of human employees in the tourism experience, how true is it of the working animals?

Defining work, non-work, and leisure

Work is, today, situated in opposition to non-work (Carr, 2017). The knowledge that once we leave school and before we retire we *must* or at least *should* work is situated in opposition to non-work. Here there is a period of our lives dominated by work, in which small periods of non-work are permissible. In this way, work becomes, in part, a measure of time, something enshrined in the idea of the working week (40 hours per week, 9 a.m. till 5 p.m., five days a week in a modern, western setting) and the retirement age. Yet, such constructions are only relatively recent, in terms of both their nature and global reach. The idea of work as we know it today is clearly a product of the northwest European Industrial Revolution (Hamilton-Smith, 1992; Juniu, 2000). This has led to the urbanization and suburbanization of the globe's population and the physical dividing of people's lives between the work and non-work space. This divide has contributed to the temporal division of work and non-work. These divides have become an integral part of how we define work and shaped the hierarchical setting in which various types of work and workers are now situated.

Work has come to be defined as what we do in order to earn the money we need in order to survive and pay for what we desire. This is, of course, a neo-liberal capitalist conception of work, but this is the dominant mode of human societies across the globe in the 21st century. Yet work is not simply about earning money in order to survive or to meet higher needs on Maslow's (1970) hierarchy of needs. Rather, it has been impregnated by humans with social and cultural value. It has resulted in the idea of work being a social good and a status symbol, with status being further enhanced by the products that working can enable us to buy. This is also clearly linked to religious constructions of a *worthwhile* life. So we see work being valued over non-work and that non-work must be earned through work (Roberts, 2006). All of this feeds into and upon itself in a continuous cycle, reaffirming what is meant by work and its dominant position in human life.

The notion of a hierarchy of work is based on the idea that some types of work, and therefore the workers engaged in this work, are more valued than others. This is clearly evidenced by the fact that people working at different tasks for the same amount of time are paid differing amounts. In this way, society places a premium value on skills, generally paying more to people with rare skills that are in high demand. Yet, it is not this simple. We see that discrimination is rife in the work hierarchy, such that equally skilled individuals are not all paid equally. Instead, there is a stubborn resistance, for example, in societies to the idea that women should be paid the same as their male colleagues, even if the gap has been declining in recent years (Lips, 2013; Tharenou, 2013). Such discrimination is not restricted to gender. In addition to valuing different types of work and workers differently, human society has so devalued some types of work as to strip them of the title of work. The obvious example here is housework, which encompasses everything from cleaning the house to taking care of dependents (e.g. children). All too often this is defined as being "not real work." That it is not monetarily compensated is both a consequence and cause of it not being defined as real work. If activities like housework are socially defined as non-work, then those engaged in it cannot be defined as workers.

It is problematic when we live in a global society that values work so highly. We define a good member of our societies as one who actively contributes to it and we place a premium on those who contribute via work. We see this notion imposed upon children from an ever-younger age, as education at school is transformed into the production of productive future workers. In the same way, retirement is defined as a social problem; what to do with a long-term unemployed segment of the population? Subsequently, we are seeing questions raised around the *right* to retire. In effect, we are asking whether people have worked enough to earn the right to retire. The answer coming back, from societies and the governments they elect, is increasingly *no*, which is why we see official retirement ages being increased.

While this book, and hence this chapter, is clearly focused on work, it is important to talk about non-work as well as it is part of how we define and see work and leisure. When we talk about non-work, at its core we are talking about all that time when we are not working. This, of course, is the time we are not engaged in that productive and socially sanctioned and approved work discussed above. In this way, non-work also refers to all those things we do because we have to but which society refuses to reward us for. So, the housework and all the roles women in particular have traditionally been associated with under the label of *duty of care* fit under the heading of non-work when talking about work as leading to social and financial rewards for services rendered.

Within the period of non-work that fills our lives lies the potential for leisure and recreation. The debate about the distinction between these two is long-running and convoluted. Anyone wishing to delve into it can wander through the leisure studies literature. For the purpose of this chapter it is sufficient to recognize that in our neo-liberal capitalist societies there is a dominant (though not necessarily accurate) notion that leisure and recreation are not a given, but rather must be earned. In childhood, good behavior and high performance at school can lead to leisure rewards (i.e., if a child does various household chores – also known as work – then they may be *allowed* to have time playing on the Xbox or PlayStation). In adulthood, we earn the right to undertake leisure or recreation and the financial means to do so via paid work. If our work is not sufficiently valued by society to gain a financial benefit or we are unemployed then we have not *earned* the right or the means to leisure (Henderson, 1990).

Another feedback loop to note here is how society has evolved since the Industrial Revolution to recognize the importance of recreation and leisure to the health of society in general and the workforce in particular (Bammel & Bussus-Bammel, 1996). Recreation in this context is fundamentally related to the idea of *re-creation* of the individual, of providing time for rest and relaxation, of the re-charging of the individual so that they are able to return to work in a heightened state of productivity (Carr & Young, 2018b). The absence of such periods of rest is clearly linked to dips in performance of employees and hence in the profit margins of businesses. Yet recreation offers more than this in our neo-liberal realities. It offers workers the opportunity to buy the very products and experiences they have been constructing while at work. In essence, recreation time now provides businesses with the opportunity to boost sales and worker productivity at the same time.

Leisure is about more than re-creation. It is not merely a product of the work experience and post-Industrial Revolution communities. Instead, it is about the *self* and the opportunity to explore and develop in a way that, in its purest sense, has nothing to do with work or the production of a more productive worker (see Carr [2017] for a more detailed discussion of the meaning of leisure). Yet leisure, as with everything else, is brought into the capitalist system. In this way, leisure, just as in the case of recreation, is seen in opposition to work, as earned free time from work. Again, just like recreation, it is seen and presented to society for its consumption, as an aid to greater productivity, and a window through which new, innovative products and experiences can be created. This talks to the pervasiveness of capitalism and the structures that drive and maintain it. Where does tourism fit into this? Despite a desire among many tourism academics to see it as something else, tourism is best viewed as a component of recreation, or leisure, as it is constructed by neo-liberal capitalism.

What is crucial to realize in all the definitions of work, non-work, leisure, and recreation is that they are constructed by humans for humans. However, living in the 21st century it is clear that most, if not all, non-human animals are defined in relation to such constructs.

Defining and critically assessing the concept of working animals

Animals have long been utilized by humans to help us achieve tasks. For example, human mobility has been enhanced by horsepower, be it riding or pulling something like a carriage. The history of dogs being used by humans as tools to aid a multitude of tasks is even longer. Indeed, they are often identified as the earliest domesticated animal (Holmberg, 2013). Used initially as a hunting tool and to enhance safety, they have since been utilized in a myriad of different ways. This has included them being used as drug and explosive detectors at, among other places, transportation nodes such as airports, sources of entertainment when raced or encouraged to fight one another, and sources of physical and emotional support. Yet it is important to recognize that a tool does not work. Rather, a tool is used by humans to enhance their productivity and/or life in general. In this way, there is no difference between an animal as tool and an electric drill, car, or weapon. All of these make it possible for the human to do a job or at least enhance their ability to do it. In addition, these tools can augment our leisure and tourism experiences. Here there is no difference between the racing dog or horse, the donkey at the beach, or the orangutan confined to the sex industry and an Xbox, vibrator, sex doll, or television.

The view of animals as tools for the benefit of humans is rooted in the conceptualization of animals as non-sentient entities that exist solely for the purpose of benefiting humanity. Such a view has certainly been the dominant one for a prolonged period of time (Carr, 2014). Thinking of animals as lacking sentience and unable to feel anything, or exhibit anything beyond automated responses to external stimuli has allowed humanity to see them as tools for our benefit and/or enjoyment. It has allowed us to utilize them in countless experiments, to eat them, and to wear their fur as a fashion accessory. We have also devised innumerable ways of using animals to entertain us. For example, they have been made to perform in circuses, ride in events such as steeple chases and dressage, and act in movies. We have also seen animals programmed to act as guides for the blind, detectors of illicit substances, and security guards.

Focusing on the tourism experience, a wide array of animals have been utilized as tools for human gain. We can think here of the animals that are used to transport tourists, including elephants, camels, horses, donkeys, dogs, llamas, and reindeer, amongst others (see Carlos-Garcia & Tallberg, Chapter 7; Hoarau-Heemstra & Nazarova, Chapter 8; Vasilopoulou, Chapter 10). Other animals are used to entertain tourists, either now or in the not-too-distant past. This includes dancing bears in Europe and black bears in the National Parks of North America who were fed by tourists, and tourism operators eager to attract tourists (Biel, 2006). Various cetaceans are used in the same way. Think of the shows offered by aquaria that house dolphins and orcas, and the myriad opportunities offered by tourism operators throughout the world to swim with cetaceans in the wild. Other entertainment opportunities are provided through the hunting, to kill or take photos of a huge range of animals, including mammals, fish, invertebrates, reptiles, amphibians, and birds (see Tully & Carr, Chapter 6; Venegas & Lopez, Chapter 4). Some animals used for human entertainment are situated within man-made spaces, where viewing them is deliberately easy. Zoos and aquariums are, of course, obvious examples. Others exist in the *wild* (itself a human-construct, and one increasingly as bounded as any zoo) (see Bueddefeld & Benbow, Chapter 12; Speiran, Chapter 11). Some tourists are simply content to see these animals, others wish to take photos of, or with them. This has only reached a new pinnacle with the advent of the digital selfie and social media rather than being anything new. Others, still, wish to have more intimate reminders of their animal experiences. These reminders span from the non-invasive to extremely invasive. They include the collection of discarded bird feathers or empty seashells, stealing eggs from the nests of birds, or taking home the head (or other body parts) of hunted animals. If animals are merely unfeeling entities put on this planet for the enjoyment and benefit of humans, then all of these are equally unproblematic. If we view animals as non-sentient beings, biological automatons, then while many are used by humans, both inside and outside of the tourism experience, none of them are actually working, as defined by human society. They are tools, not workers or employees.

Well-being and the working animal

If we view animals as non-sentient, biological automatons then do we need to be concerned about the well-being of those we, as humans, use as tools? The answer clearly is yes, and society has long recognized this. Where animals are used as tools it has long been recognized that in order to gain maximum benefit out of them they need to be treat well. This means not only providing an animal with appropriate food and water, but also with sufficient non-work time to enable it to *re-create*. In this way there has long been a concern with providing animals time off from work, just as provided for human employees. This has been enshrined in law. Failure to allocate this time off work, at minimum, can damage a valuable tool. We seek to maintain the cars we own by giving them regular services and ensuring we put good quality oil and petrol in them. We may even store them inside when we are not using them to guard against damage from adverse weather or from them being stolen because they are valuable. In the same way, a valuable animal is carefully treated and maintained. Think, for example, of the resources that go into the maintenance of the best racehorses or show dogs in the world.

When we think of the well-being of so-called non-sentient animals, we find society deciding how important it is based on the value of the animal as a tool. So, to take the example of the top quality racehorse further, we see breeders and owners (recognizing tools are owned and that in human law animals are owned, something that is a throwback to the societal construction of animals as objects, non-sentient beings) of these horses lavishing attention on their well-being. However, once their ability to race to a high standard (as defined by humans) is at an end, either due to age or injury, and if they have no breeding potential concern about the well-being of the horse rapidly declines in line with its declining value (Dashper, 2014, 2017). The end state of this process is the horse being sent to the abattoir (Winter & Young, 2015). Throughout all of this, the focus is never on the well-being of the animal for the benefit of the animal. Rather, the focus on well-being is for the benefit of the owner.

Yet, increasingly, humanity is recognizing that animals are sentient (Bradshaw, 2011; Goodall, 2007). In recognizing this we are seeing a shift away from the idea that non-human animals are merely objects and that any feelings they experience (pain, fear, joy, etc.) are nothing more than simple autonomic responses. In this way, we are bridging the divide between animals and humans, suggesting that if we believe such responses and feelings are more than simply unthinking, autonomic responses in humans then we can no longer assume the same is not the case in animals.

However, the idea of animal sentience is not universally agreed upon. Instead, arguments continue to rage about whether such sentience is comparable to human sentience and if it extends, and to what extent, to all species. There is also an ongoing tension between natural scientists and social scientists exploring sentience. The former cling to the foundations of modern science that demand testing and the provision of observable proof. As a result, even among those who concur with the notion of animal sentience, this conclusion is grounded on ideas of stimulus and response. In this way, such scientists do not consider the conscious ability of animals to think and feel beyond what they can see. In contrast, social scientists, particularly those who have distanced themselves from the traditional positivistic ideals of science, tend to look beyond the surface, observable facts. Here we begin to talk not just of sentience, but also of consciousness, emotions, and, without wishing to get into a religious discussion, the soul. These are the things we take for granted in humans (even the soul, irrespective of our religious beliefs, though we may talk and think of it in different terms). It is the idea that we are more than merely a collection of biological parts put together in a neat way that enables us to perform a lot of complex tricks that are far beyond the abilities of any computers invented to date.

Yet despite, in a scientific sense, not being able to "see" consciousness or emotions, more and more people are subscribing to the idea that they are not restricted to humans. Instead, we have seen the development of the idea of the conscious, sentient animal spreading across species in recent years. The idea first began to emerge in relation to specific species, those identified by science and social desires as being most like us. So, sentience was first ascribed to the great apes, before spreading to porpoises and then supposedly less intelligent animals such as dogs and horses. Today, there is increasing evidence for, and belief that we should just recognize that all animals are sentient (Carr, 2014; MacFarland & Hediger, 2009). Such a step is one of faith at the moment as research has yet to provide conclusive proof to support it. It means overturning the idea that something cannot be true until it is totally proven. Yet such is the weight of evidence now available that we are not talking about a complete reversal of scientific thinking. Rather, it can be seen as a rational leap of faith, and increasingly just a small step.

If we see animals as sentient beings, we cannot regard them as tools. Instead, we must recognize them as workers. They may still, on the surface, be undertaking the same task, but the underlying implications are significant. Once we recognize animals as workers, we must recognize that while non-work time, incorporating recreation and leisure, for them may still benefit humans there are also important potential benefits to the animals. Furthermore, as workers, in alignment with human constructions of work and workers, the benefits to the animals must take precedence over the benefits to humans (for a detailed discussion of the work of Camilla, a captive chameleon living in a theme park in the UK, readers are encouraged to look at Wilkinson [2018]).

Yet, relatively few people give conscious thought to the way we increasingly talk of animals as workers or even employees. The use of such verbs is partly an anthropomorphic kneejerk (ironically, unconscious) reaction, that enables us to think of animals in human terms. This is far easier to do than thinking of animals in animal terms when humanity has, for so long, worked to define itself as something distinct from all the other animals on the planet. In the tourism experience we arguably give even less thought to the animal as worker, seeing everything in the experience as being part of the entertainment we have bought through our own hard work; in other words, something we have earned. However, just as thinking of humans working in the tourism experience in this way is wrong, if we believe in the sentience of animals, it is also wrong to think this way in relation to working animals. In this context it is important to remember that even animals in the entertainment industry are workers.

The implications of thinking of animals as workers are substantial. If they are workers and not slaves, the distinction being that the former while needing to work are at least free to walk away from employment and are paid while in a job, then we cannot *force* them to do something. Forcing a sentient animal to do a job would be slavery. Thinking in this way means that no animal should be forced to work, and that they should all have the right to say *no* to working. In this way, the lion in the zoo that does not wish to take part in an educational display should not have to. Neither should the greyhound have to race if it does not feel like it or the husky pull a sled if it does not feel inclined to do so. In addition, we must consider how, when at work, an animal's well-being as a sentient being is taken into consideration. Yes, there are already in place many laws and regulations that are designed to ensure the welfare of animals, but these are not focused on animal's themselves. Rather, they are based on the idea of the animal as human property. Alternatively, these laws are designed to ensure human feelings of guilt about perceived mistreatment of animals are assuaged. These are human-centric rather than animal-centric constructions. They may enhance animal welfare, but they are not grounded in animal-centric thinking.

Yet, work in a human context is an exchange, albeit it often an unequal one, where the worker's time is exchanged for, generally in the 21st century, money. If animals are workers, then we must ask, "what do they gain in return?" Is it that the detection dogs on patrol in airports are *paid* in food and shelter in return for their skilled work? This could be said to be the case, but equally we could say that these things are provided to maintain the dog at peak performance. The food and shelter may be the same but the intention behind the act of giving them matters. Looking beyond the working life of an animal, what happens when they retire? When a human retires, they will have either saved enough in a state or private pension or hope they can be cared for by the government or relatives. What they can be sure of is that they will not be taken to the abattoir and disposed of. As mentioned earlier, when a racehorse or a greyhound, for example, is retired they may be deemed surplus to requirement and then they are at the mercy of a society that may voluntarily seek to rehome them or turn a blind eye to their slaughter. If we perceive animals to be working, then do we need to consider that they may have a need or desire for play, recreation, and/or leisure in addition to having downtime from work and annual holidays. This was a core component of the work highlighted in the edited book by Carr and Young (2018).

Another question to consider is whether imposing the idea of work in exchange for something (such as food and shelter, but also the *right* to continue to survive) is in itself a form of animal abuse. Whale watching is an excellent example of this. The argument has been made that whale watching is an economic justification for not engaging in whale hunting (Wilson & Tisdell, 2003). Following this logic, the whales are required to work as tourist attractions in exchange for not being driven to extinction. This is not a rationale for work that would be defined as socially acceptable if it involved humans. Recognizing animals as sentient beings means that we must question the acceptability of this rationale when applied to whales or any other animals.

Animal-centric thinking

In this chapter, it has been argued that if we see animals as sentient beings, different but related to the sentience we recognize in humans, then we need to see animals as workers rather than tools. Following this logic, we recognize the need for animal welfare in the workplace to be animal-centric rather than defined by humans for the benefits of humans. This conforms to Broom's (2010) conceptualization of the obligations humans have toward animals. It recognizes that humans, including, but not restricted to, the employers of animals and consumers of the products and experiences that animal workers provide, are in a position of power over animals. The responsibility, or obligation, of these people is to utilize this power to enable the welfare of animals rather than to reinforce their abuse.

However, we need to step back from the argument put forward so far in this chapter to examine whether we are, once again, guilty of assuming a human-centric perspective. To do this, we must consider whether animals themselves see what they do when working as *work* in the same way as humans do. In the same way, we must consider whether animals see the divide between work and non-work in the same way humans do and if they need or desire play, recreation, and leisure as humans do. In order to consider this, the myriad working dogs can be used as an example. Sniffer dogs and racing dogs are examples where what they do naturally and what they appear to thoroughly enjoy (incorporating ideas of fun and pleasure) doing is harnessed for the benefit of humans. Herein lies the potential for animal abuse due to power differentials between humans (those in power) and animals (the powerless). It is important to consider whether the dogs, in this case, see the work (as defined from a human-centric perspective) as work. However, just because animals or humans enjoy doing something does not necessarily mean it is not work. Rather, we may need to think about whether animals feel obligated or forced to undertake the activity. Moreover, do the animals that we define today as working feel the need for non-work and the leisure and recreation that is an integral part of the non-work ideals of humans. It is beyond the limits of this chapter to consider whether animals seek, need, and/or desire leisure, recreation, or a holiday but it is certainly the case that humans have anthropomorphized various animals to consider them to be taking holidays at various times of year. An example is the pit ponies who used to be *employed* down mines pulling coal out of the ground. Their break from this work in the dark during a summer period when they were able to feel the sun on their backs and frolic in fields like horses are *meant* to do has been depicted as their holiday. Yet, a look at Moist Von Lipwig's struggles in Terry

Pratchett's (2013, p. 161) *Discworld* with a golem horse shows that the provisioning of *holidays* or *leisure* for horses may be more for the benefit of humans than animals. Moist starts:

> "Would you like to run around in meadows and generally cavort in pastures and so on?"
> Out of nowhere came [the response from the golem horse], "Yes, if you wish."
> Moist said, "But what do *you* wish?"
> "I don't understand the concept."
> Moist breathed in and said, "I saw a little stream not far back, and some green pastures and, for the sake of my soul, I would like you to go over there and gallop in the meadows and enjoy yourself."
> "Yes, enjoy myself, if you want me to."
> "For heaven's sake, this is manumission we're talking about here!"
> "That would be horseumission, sir. And I must point out that I don't need to enjoy myself."
> "Well, do so for my sake, will you, please? Roll around on the flowers and neigh a bit and gallop about and have some kind of fun. If not for your own pleasure, then for my sanity, please."

This all brings us back to the need for those in power (i.e., humans) to recognize and act on their obligations to the animals in their care (an important distinction from being owned). Take the example of my dog, Ebony. If she ever works, it is as an emotional support for me and the rest of our family. Is leisure for her running at the beach and chasing tennis balls? Certainly she gives all the indications that she loves these activities and will happily keep returning the ball to me to throw again for hours on end. Yet here is the important part, I have to be careful to take her home before too long or else she will, literally, keep running herself into the ground in pursuit of her tennis ball. So, I have as the person in power (of deciding to throw the ball or not, though she will voraciously demand the ball be thrown again if I hold it in my hand too long) have the obligation to ignore her demands that the ball be thrown again and instead take her home to lie down after a long drink of fresh, rather than salt, water. This is the exercising of power to meet the obligations of animal welfare.

If we consider animals to be sentient beings, we must ask whether we have the right to determine what animals can and cannot do, and as a result whether they have the right to say they should or should not work in the tourism industry. If an animal is happy and engaging in the work at least as voluntarily as humans engage in work, then do we have any right to prevent them engaging in this work. This steps into the emotive divide between animal welfare proponents, of which I am one, and animal rights proponents who, at their most extreme, do not think animals should be used to benefit humans in any way (Rudy, 2011). Such a stance arguably removes the right of the animal to choose. The problem is that the animal must clearly be capable of and empowered to make the choice. For the racehorse, even if it were capable of rational thought about the issue of racing or failing to race and being turned into meat the choice is not a fair one. Once again, the outcome is a recognition of the importance of human obligations, in this case to make sure that animals have reasonable alternatives. In the case of the racehorse the outcome is that those who can or no longer wish to race do not go straight to the abattoir but have a post-racing life to enjoy instead.

Conclusion

So how should we think about animals working in the tourism industry, or any other context? The important thing is not to worry about whether we, as humans, think an animal is working or not. After all, this is only human semantics that animals would not understand or appreciate. Instead, the need is to focus on the welfare and well-being of the working animals. Utilizing semantics can certainly help to encourage thinking about animals as sentient beings and hence their welfare needs, but it can only take us so far. We should not, in the rush to think of animals as sentient beings, fall into the trap of thinking of them as the same as us and then foisting human constructs upon them. Instead, if we think of animals as working in the tourism experience, we need to think of that *work* in animal terms, not human ones. Following on from that is the need to think of breaks from work, non-work, play, leisure, and recreation from an animal perspective. The biggest trick though is to think about working animals from both animal and human perspectives at the same time. This is necessary when recognizing that although animals are sentient beings they still live within human bounded realities.

References

Bammel, G. & Bussus-Bammel, L, (1996). *Leisure & human behavior* (3rd ed). Madison: Brown & Benchmark Publishers.

Biel, A. W. (2006). *Do (not) feed the bears: The fitful history of wildlife and tourists in Yellowstone*. Lawrence: University Press of Kansas.

Bradshaw, J. (2011). *In defence of dogs: Why dogs need our understanding*. London: Allen Lane.

Broom, D.M. (2010). Cognitive ability and awareness in domestic animals and decisions about obligations to animals. *Applied Animal Behaviour Science*, *126*, 1–11.

Carr, N. (2014). *Dogs in the leisure experience*. Wallingford, UK: CABI.

Carr, N. (2017). Re-thinking the relation between leisure and freedom. *Annals of Leisure Research*, *20*(2), 137–151.

Carr, N. & Broom, D. (2018). *Tourism and animal welfare*. Wallingford, UK: CABI.

Carr, N. & Young, J. (2018). Wild animals and leisure: An introduction. In N. Carr & J. Young (Eds). *Wild animals and leisure: Rights and welfare* (pp. 1–11). Abingdon: Routledge.

Carr, N. & Young, J. (2018b). Conclusions: Charting a way forward. In N. Carr & J. Young (Eds). *Wild animals and leisure: Rights and welfare* (pp. 225–233). Abingdon: Routledge.

Clarke, J. & C. Critcher. (1985). *The devil makes work: Leisure in capitalist Britain*. Basingstoke: MacMillan.
Dashper, K. (2014). Tools of the trade or part of the family? Horses in competitive equestrian sport. *Society & Animals*, *22*(4), 352–371.
Dashper, K. (2017). *Human-animal relationships in equestrian sport and leisure*. Abingdon: Routledge.
Goodall, J. (2007). Foreword. In M. Bekoff (ed.). *The emotional lives of animals* (pp. xi–xv). Novato, California: New World Library.
Hamilton-Smith, E. (1992). Work, leisure and optimal experience. *Leisure Studies*, *11*(3), 243–256.
Hart, L. (2003). Pets along a continuum: Response to "What is a pet?" *Anthrozoös*, *16*(2), 118–122.
Henderson, K. (1990). The meaning of leisure for women: An integrative review of the research. *Journal of Leisure Research*, *22*(3), 228–243.
Holmberg, T. (2013). Trans-species urban politics: Stories from a beach. *Space and Culture*, *16*(1), 28–42.
Juniu, S. (2000). Downshifting: Regaining the essence of leisure. *Journal of Leisure Research*, *32*(1), 69–73.
Lips, H. (2013). Acknowledging discrimination as a key to the gender pay gap. *Sex Role*, 68, 223–230.
Maslow, A. (1970). *Motivation and personality* (2nd ed). New York: Harper & Row Publishers.
MacFarland, S. & Hediger, R. (2009). Approaching the agency of other animals: An introduction. In S. MacFarland & R. Hediger (Eds) *Animals and Agency: An Interdisciplinary Exploration* (pp. 1–20). Boston, Massachusetts: Brill.
Page, S. & Connell, J. (2010). *Leisure: An introduction*. Harlow: Prentice Hall.
Pratchett, T. (2013). *Raising Steam*. London: Doubleday.
Roberts, K. (2006). *Leisure in contemporary society* (2nd ed). Wallingford, UK: CABI.
Rudy, K. (2011). *Loving animals: Toward a new animal advocacy*. Minneapolis: University of Minnesota Press.
Tharenou, P. (2013). The work of feminists is not yet done: The gender pay gap – a stubborn anachronism. *Sex Roles*, *68*, 198–206.
Wilkinson, S. (2018). Being Camilla: the 'leisure' life of a captive chameleon. In N. Carr & J. Young (Eds). *Wild animals and leisure: Rights and welfare* (pp. 96–112). Abingdon: Routledge.
Wilson, C. & Tisdell, C. (2003). Conservation and economic benefits of wildlife-based marine tourism: Sea turtles and whales as case studies. *Human Dimensions of Wildlife*, *8*(1), 49–58.
Winter, C. & Young, W. (2015). Fatalities and fascinators: A new perspective on thoroughbred racing. In N, Carr (Ed). *Domestic animals and leisure* (pp. 241–258). Basingstoke, UK: Palgrave Macmillan.
Young, J. & Carr, N. (2018). Domestic animals leisure, rights, wellbeing: Nuancing 'domestic', asymmetries and into the future. In J. Young & N. Carr (Eds). *Domestic animals, humans, and leisure: Rights, welfare, and wellbeing* (pp. 209–222). Abingdon: Routledge.

Part 2: **Performative work**

Gino Jafet Quintero Venegas, Álvaro López-López

4 Working donkeys in northwestern Mexico: Urban identity and tourism resources

Introduction

For over a hundred years, the “zonkey” (zebra-donkey) has been a tourist icon of the border city of Tijuana (Stolk, 2004). The name of this creature, whose job is to pull a cart to carry tourists along the streets, stems from the black stripes painted on it to simulate a zebra. Tourists usually wear a colorful poncho and a traditional hat for the photo opportunity. Regarded as heritage and a symbol of the city’s identity, this non-human animal has working days of over 12 hours (Caballero, 2014). It does not matter that zebras are not part of Mexico’s natural fauna: this banal and decontextualized image works very well for marketing purposes, and its fictitious character reinforces its recreational value (Cohen, 2005).

The very existence of the zonkey is controversial from the perspective of animal rights and a critique of anthropocentrism. Not only is its well-being a concern, but its recreational use is problematic (Fennell, 2012; Horta, 2017). When a donkey is used to entertain tourists, the latter reinforce their position of superiority, their dubious “right” to unlimited individual happiness, with no regard for bioethical concerns (Skolimowski, 2017). When the attempt was made to declare the zonkey a heritage element of the city, activists protested and tried to revoke the initiative (Uniradio Informa, 2017).

The purpose of this chapter is to explore the historical context in which a city’s identity has been built around the image of this sentient being. In order to question the ethical legitimacy of naming a work animal as cultural heritage, tourism practices in Tijuana are subjected to scrutiny through the following questions: Is it ethically legitimate to sacrifice the well-being of a sentient being for the benefit of a city through tourism? To what extent can a living creature be commodified as a tourist attraction and a heritage element?

This chapter draws form the theoretical corpus of the “third wave” of animal geography (Urbanik, 2012). Based on a post-humanist positioning, a triangulation of several qualitative techniques will be attempted (Buller, 2015; Decrop, 1999). A thorough revision of the literature and a documentary search were carried out in both printed and digital formats on animal ethics, work animals, time-space dimensions of tourism development in Tijuana, the consolidation of the image of the

Acknowledgments: The cartography included in this chapter was elaborated by Lourdes Godínez Calderón.

https://doi.org/10.1515/9783110664058-004

zonkey as a tourism resource, and cultural heritage. Likewise, tourists' discourses on the zonkey on social media platforms were analyzed (Xiao & Zhao, 2009).

The chapter has been organized into five sections. In the following section, humanism and post-humanism, the two philosophical perspectives that legitimize contemporary tourism practices with work animals are analyzed. Next, the ethical postulates for and against the use of non-human animals for work are contrasted. Then, a historical overview of the development of tourism in Tijuana and the adoption of the zonkey as a recreational resource is presented. Here, the exploitation and abuse inflicted on the zonkey and the ethical legitimacy of turning the zonkey into a heritage resource are also discussed. Finally, the chapter concludes by stating that the commodification of the zonkey as a tourist resource and heritage is the product of the unethical supremacist ideology of speciesism. The section closes with a call for the immediate abolition of practices that force animals to work without their consent.

Tourism and work animals: Humanist and post-humanist perspectives

Non-human animals are and have been systematically used as workforce and their use has been naturalized without questioning the pertinence of their activities. Deprived of their freedom and basic rights, they are commodified, their behavior conditioned, and they are appreciated only in terms of their extrinsic value for the benefits they provide (Lynn, op. cit.; Taylor & Signal, 2009). Arbitrary roles have been assigned to them in current economic systems, including the construction of tourism environments (Markwell, 2015).

Urbanik (2012) establishes three categories for work animals: animals for education (those destined to labs and zoos), service animals (for transportation and traction, guardians, and guides), and animals for recreation (used in film studios, circuses, fights or tourist attractions). Work can be exhausting and can cause them pain or psychological damage (Swann, 2006). Their exerting and monotonous routine, often accompanied by physical or psychological punishment, is a source of stress, fear, and anxiety. These beings live very unhappy lives and often die of exhaustion (Hribal, 2003).

Furthermore, Hribal (2003) claims that non-human animals are part of the working class. Being unreliable workers, he argues, they partly prompted the Industrial Revolution and were a driving force in the rise of capitalism. Their cooperation and resistance also shaped human labor and instruments and, following this logic, non-human animals have been used as one of the main workforces that ensure the success of the tourism industry worldwide.

The generalized idea that our own interests are above our moral obligations towards other animal species has led to their commodification. This false ethical hierarchy has a serious impact on the individual and collective well-being of non-human animals. Since the second half of the last century, academic thought has challenged the discriminatory position towards other animal species (Horta, 2017), an ideology branded as "speciesism" by Richard Ryder in 1970 (Singer, 2009).

Western philosophical thought has reproduced two fallacies responsible for the maltreatment of non-human animals: that they are inferior, and therefore at our service; and that they are not sentient beings as humans are, and therefore they are replaceable automatons (Cottingham, 1978). Moreover, the modern humanist paradigm maintains that human animals are above all other forms of nature, and that the ultimate purpose of all-natural elements is to satisfy human needs (Lynn, 1998). Thus, any practice in which human animals are used as simple means of production has been normalized (Weitzenfeld & Joy, 2014).

Post-humanism, a philosophical paradigm that emerged at the end of the 20th century (Harari, 2016), pushes for a more inclusive ethical pluralism to account for other animal species and for a revision of the idea that reasoning is an exclusively human attribute (Urbanik & Johnston, 2017; Wolfe, 2010). By deconstructing anthropocentric human-non-human relations, post-humanism seeks to transform the ethical grounds of tourism and recreational experiences and proposes a qualitative move from animal ontology to animal rights. Post-humanism questions and seeks to abolish those activities in which non-human animals are perceived as having only a utilitarian value (Urbanik, 2012; Wolfe, 2010). Though this paradigm has a growing presence in the social sciences and humanities, speciesism is still accepted in texts on tourism studies. It seems that humanism, firmly entrenched as it is in the collective imaginary of societies.

Ethical grounds for the use of non-human work animals in tourism

In the 1980s, the sustained increase in the number of tourists worldwide lead to the privatization of air transport and the opening of new global routes; the international hospitality industry was consolidated and expanded, and new tourist destinations were promoted, such as beaches in the subtropical littoral of Latin America and Southeast Asian cities (Walton, 1997). As a result of this impulse, the UNWTO General Assembly called for a meeting in Istanbul, Turkey, in 1997 to create a code of ethics for tourism to be used by local residents, tourism operators, and visitors in order to identify legitimate tourist practices (Fennell & Malloy, 2007).

The WTO's "Global Code of Ethics for Tourism" is a reference for the development of responsible and sustainable global tourism, in tune with the ideas about human society prevalent in the 21st century (Organización Mundial del Turismo [OMT], 1999). It was convened by more than 60 countries, with the conviction that tourism can significantly contribute to improve both the quality of life of human beings and the ecosystem (OMT, 1999). Directed to governments, companies, communities, and tourists alike, the document sets a standard for the acceptable practices of both tourists and service providers, with an end to deflect tourism's negative impact on the environment, cultural heritage, and social life (Upchurch, 2000; OMT, 1999). However, the document's ethical positioning is both anthropocentric and instrumental, and its application is not compulsory. Economic benefit being the ultimate goal, it fails to condemn the use and exploitation of non-human animals as a resource or heritage element:

> Tourism activities should be conducted in harmony with the attributes and traditions of the host regions and countries and in respect for their laws, practices and customs.
> (Article 1, OMT, 1999)

> Tourism, [. . .] and access to culture and nature, should be planned and practiced as a privileged means of individual and collective fulfilment. . . (Article 2, OMT, 1999)

> Tourism activity should be planned in such a way as to allow traditional cultural products, crafts and folklore to survive and flourish, rather than causing them to degenerate and become standardized. (Article 4, OMT, 1999)

Article 2 of the Code, in its sections 2 and 3, talks about the values that must be fostered within tourism – including respect for human rights and vulnerable groups – and states that those who engage in exploitative practices should be sanctioned. All these rules, however, belong in an anthropocentric framework:

> Tourism activities should respect the equality of men and women; they should promote human rights and, more particularly, the individual rights of the most vulnerable groups, notably children, the elderly, the handicapped, ethnic minorities and indigenous peoples. (OMT, 1999)

> The exploitation of human beings in any form, particularly sexual, especially when applied to children, conflicts with the fundamental aims of tourism and is the negation of tourism; as such, in accordance with international law, it should be energetically combatted with the cooperation of all the States concerned and penalized without concession by the national legislation of both the countries visited and the countries of the perpetrators of these acts, even when they are carried out abroad. (OMT, 1999)

In order to combat the exploitation of non-human animals in tourism, the document's moral spectrum should at least widen so as to include and begin to consider them as sentient beings with their own interests. If tourism practices exploitative of non-human animals have been socially accepted it is because the animals are considered to be far from *homo sapiens* and thus outside of moral considerations (Fennell, 2012). But from an anti-"speciesist" standing, they are the most vulnerable social group on the planet. A simple rewriting of the aforementioned article in the Code would suffice to reverse this situation:

> [. . .] they [tourism activities] should promote human rights and, more particularly, the individual rights of the most vulnerable groups, notably children, the elderly, the handicapped, ethnic minorities [. . .] indigenous peoples [, and non-human animals].

Based on this directive, all practices bent on assaulting the primary interests and well-being of non-human animals could be condemned and prohibited.

Today, many tourism practices that perpetrate animal abuse are justified by "welfarism" (Quintero & López, 2018). This ethical stand maintains that owning and using animals is morally acceptable as long as unnecessary suffering is avoided or reduced (Mendl & Paul, 2004). Opposed to welfarism are other ethical positions that give non-human animals the same status as human animals in terms of avoiding suffering and preserving life. These tenets have not been well received in academic circles and in civil society in general, as they challenge anthropocentrism and the planetary superiority of humans (Urbanik, 2012).

One of them is ecocentrism, which argues that individual rational actions and thoughts must be directed above all towards caring for and conserving the environment (Lynn, 1998; Fennell & Nowaczek, 2010). Ecocentrism is important, for it challenges the separation of humans from nature and thinks of the ecosphere as an inclusive, complex, integrated, creative, beautiful, and mysterious whole, older than time itself (Lynn, 1998). Similarly, Regan's animal rights theory (1987) purports that every being, human and non-human, has an intrinsic value that must be considered from a position of responsibility and compassion. Based on respect and egalitarian justice, it rejects any moral hierarchies between individuals and species, and denounces most of the recreational and touristic activities that use non-human animals.

Furthermore, utilitarianism maintains that human actions are morally good if they maximize happiness and well-being and avoid pain or suffering for the majority of the individuals involved. Clearly, in the use of non-human animals to satisfy the desires of tourists, there is a conflict of interests, as this is done at the expense of the happiness and well-being of non-human animals. Perceived as they are as resources for the creation of touristic experiences, they are often exposed to unethical practices. This is only possible in the context of domination, violence, and disregard for otherness that pervades human and non-human animal relations.

In their analysis of the existing power relations in the realm of tourism (between tourists and service providers, local inhabitants, industry and institutions, among other actors), Cheong and Miller (2000) and Bramwell and Meyer (2007) fail to include the last link in the chain of moral considerations: non-human animals, who bear the brunt of this dynamics, and are systematically exploited, objectified and rendered invisible (Markwell, 2015). Based on the argument that animals are also sentient beings, capable of feeling pain and pleasure, and that they have an interest in preserving life, Singer (1975) questions the existing relations of domination that humans have established over non-humans for their own benefit, and states that non-humans should not be treated as resources.

Another school of thought that has contributed to this discussion is ecofeminism. Studies on ecofeminism have focused on the links between the oppression of women and the oppression of non-human animals, both a product of the same logic of domination. The central argument is that nature (and other animals) cannot be liberated until the oppression of women is taken seriously, and vice versa (Adams & Gruen, 2014). The deconstruction of speciesistic paradigms has shown that women's subordination and the domination of nature are the two sides of the same coin, responding to the common logic of patriarchal capitalism and the subordination of life to economic gain (Christ, 2006).

The adoption of the zonkey as a recreational and heritage resource

With 2.1 million inhabitants, Tijuana is the fifth most populated city in Mexico (INEGI, 2018). Along with the neighboring city of San Diego, it forms an international metropolitan area spreading on both sides of the border. Tijuana's growth is a result of the establishment of assembly plants (*maquiladoras*) in the city in the 1970s (Walker, 2011). Its border location and loose geographic articulation with central Mexico makes its tourism activity almost entirely dependent on visitors from the United States (Carmona & Correa, 2008) (Figure 4.1).

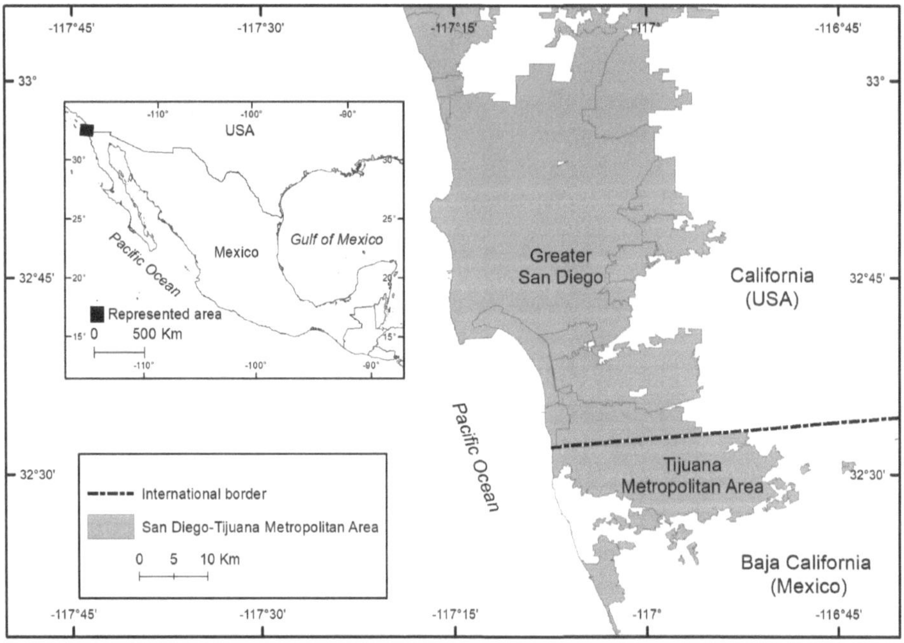

Figure 4.1: Location of Tijuana in northeastern Mexico.
Source: Lourdes Godinez. Used with permission.

The story of the zonkey dates back to the end of the 19th century, when the image of Mexico was similar to that of the Wild West in the imaginary of the US public. As soon as travelers from the US crossed the border into Mexico, they were offered to take a photograph to capture their stereotypical tourist experience: mounting a donkey and wearing traditional Mexican garb. However, since the images were far from sharp, the donkey was lost between the black-and-white tonalities (Cháidez, 2012).

In the 20th century, Tijuana was consolidated as the central urban node in the Tijuana-San Diego tourism system. The city began to receive large contingents of visitors as alcohol-related businesses moved across the border after the proclamation of the Dry Laws (California, 1915; Texas, 1918) (Carmona, 2003). As a result, US authorities began to control the flow of visitors and limited the crossing times to before 6pm in order to "safeguard both the moral and the social condition of the San Diego population" (Carmona, 2003, p. 47). Although these actions negatively impacted the local economy, the hotel sector benefited from them, as spending the night on the other side of the border became necessary (Ceballos, 2001; López & Bringas, 2016).

In order to make the city more attractive to foreign visitors, the recreational offer was expanded to include betting games and shows with non-human animals (Carmona, 2003). In 1927, the Agua Caliente enclave was built three kilometers away from the city center, featuring a casino, a greyhound-racing course, and a horse-racing course, and opening new jobs for local residents (Cháidez, 2012). With the increasing arrivals of US tourists to the city, the local government institutions began to consolidate the zonkey as an icon of the city. In 1938, betting games were prohibited in Mexico (Andrade, Moreno, & Quiñones, 2015). When the Agua Caliente casino closed down, the habit of taking a snap of the donkeys had become so ingrained, that the offer multiplied throughout the city. The *burreros* (donkey holders) had moved to the Avenida Revolución, the axis of tourist mobility leading to the borderline (Carmona, 2003).

The oldest zonkey picture known dates from 1939, when the non-human animal was already located on Avenida Revolución; however, no image record of this picture is available (Vincent, 2018). Cháidez (2012) says that the idea of painting stripes on the donkeys originated in a visit that a *burrero* made to the San Diego Zoo. Observing the zebras, he decided that the pattern would look good in black-and-white pictures.

During World War II, the arrival of US soldiers and marines to the military port of San Diego and to the border increased the demand for recreational spaces in Tijuana (Carmona, 2003). In order to attract more tourists, the donkey's decoration changed to include elements of the imaginary about Mexico, such as desert landscapes and fields covered by *nopal* (edible cactus) (Observatorio Turístico de Baja California, 2015). In 1945, the popularity of zonkeys was such that on Avenida Revolución there were up to 25 of these non-human animals. A *burreros* union was formed to stave off competition and ensure that only registered members could work as official photographers (Vincent, 2018).

In the 1950s, Tijuana began to acquire notoriety as a national tourist destination due to the construction of roads linking the city to the country's interior, as well as the opening of commercial air-travel routes, such as the Mexico City-Tijuana and the Acapulco-Tijuana routes (Carmona, 2003). Hybrid decoration elements were added to the zonkey, such as saddles and palm hats with legends like "Just Married," "Amigo," "Tijuana," or "The Boss" (Vincent, 2018). These new decorations were intended to make the zonkey more attractive to tourists, while reproducing the idealized folkloric *kitsch* (Palaversich, 2002).

In 1964, with the end of the Bracero program, which granted agricultural work permits to Mexicans, many migrant workers returned to the country. Tijuana's population increased and many people looked for a job in the foreign-owned *maquiladoras* (Carrillo & Hualde, 2000). However, some of those who did not find a job in the industry, turned to self-employment on the Avenida Revolución, either as informal sellers of handicrafts and other products made in China, or as unregistered *burreros* (Cháidez, *op. cit.*). Many *burreros* kept their irregular trade until the last decade of the 20th century, when the city's economy improved significantly. On the one hand, the Paisano program (the counterpart of the Bracero program) offered economic support to returning migrant workers. On the other hand, Tijuana began to be promoted as an international tourism destination (Ruiz, Martínez, & Verján, 2015). The National Tourism Program launched both road-trip tourism and hunting tourism in natural protected areas. The image of border cities improved significantly and visitors' expenditures grew substantially (Carmona, 2003).

Given its position as a node of bi-national transit, Tijuana's tourism dynamics differs from the one found in other Mexican cities. Foreign tourists seek activities that are not linked to the natural environment, but to alcohol and drug consumption, games, medical services, and prostitution (López & Bringas, 2016). It was this unique spatial context that turned the zonkey into a tourist attraction.

In 2013, the proposition came to declare the zonkey as cultural heritage of the city, given its importance in building an identity for Tijuana (Notimex, 2014). On August 8, 2014, the zonkey was officially named Cultural Heritage of the State of Baja California (Observatorio Turístico de Baja California, 2015), causing much controversy. On the one hand, those who supported the decision alluded to its cultural value; on the other hand, those who rejected it cited ethical concerns. After a number of demonstrations, in 2017 animal rights lawyers obtained a writ (*amparo*) to prevent that the zonkey be considered heritage (Newsweek Mexico, 2017).

The economic value of Tijuana's zonkey is twofold: it is a work animal and a tourist attraction (Markwell, 2015). Donkeys have long been appreciated as part of the local identity and cultural heritage. Their physical presence in the city, as well as their visual artistic representations and their promotion by the government have made them also part of the touristic imaginary (Figure 4.2). It is not hard to understand that tourists want to experience contact with them and take home the traditional photograph as a souvenir (Comité de Turismo y Convenciones de Tijuana [COTUCO], 2012).

Figure 4.2: An urban image of the zonkey in Tijuana.
Source: Lapiztola, 2010. Used with permission.

The exploitation of the zonkey

The Cambridge Declaration of Conscience of 2012 concludes that non-human animals are sentient beings and conscious of what happens to them (Horta, 2017). They are subjects with will, intentions, desires, and their own interests, and they experience their lives through sensations and emotions. Life for them is an end in itself and possesses an inherent value, not a simple means to somebody else's ends. Tourist practices reinforce the existing power relations between species. Humans coerce non humans to do something in return for food, shelter, and to avoid being punished. Non-human animals have the notion that they are subjected and act out of fear, motivated by the pain inflicted on them. They consent to their enslavement to avoid retaliation. They work coerced by their exploiters and because they have no option (Francione, 2009).

The raising, trade, and consumption of animals are regarded as positive, unless extreme suffering is inflicted on them, such as keeping them in reduced spaces and exposed to the natural elements, for example, or being subjected to working routines longer than 14 hours. The use of donkeys in Tijuana is clearly a form of exploitation because donkeys are forced into a kind of work that goes against their preservation interests and can bring them suffering:

> Withholding conditions or commodities for which an animal shows "inelastic demand" (i.e., for which it continues to work despite increasing costs) is very likely to cause suffering.
> (Dawkins, 1990, p. 1)

When animal abuse in tourism practices becomes visible and is condemned, the exploiters get away with it by implementing welfarist alternatives that reduce the abuse without eliminating it. Animal welfare is based on five different rules which allows the exploitation of animals if these needs are covered properly: a suitable environment; a suitable diet; exhibit normal behavior patterns; be housed with or apart from other animals; and protection from pain, suffering, and disease (Webster, 2001).In the particular case of the zonkey, the pressure by animal rights activists led to the promulgation of norms that dictate the "correct" form of donkey exploitation, such as giving them breaks. But, far from solving the problem, these norms actually reinforce animal exploitation for economic gain (Horta, 2017).

From an abolitionist standpoint, the very concept of abuse is inapt because it exists only in opposition to "fair treatment." The notion of abuse or maltreatment does not condemn anthropocentrism or the hierarchical power relations to which non-humans are subjected. Once the moral spectrum has been widened to include non-humans, their use without consent becomes unethical (Fennell, 2012). For consent to be considered ethical, it must comply with the following prerequisites: 1) equality between the parts, ensuring that no abuse can take place; 2) total freedom to choose, with no coercion or pressure; 3) full understanding of the situation and knowledge of what is being done and what it entails (Tovar, 2015). Since none of these prerequisites can be fulfilled between humans and non-humans, any form of use of the latter by the former is a form of abuse. Animals have not given their consent to be used, have not decided by themselves, and have been submitted by force to hedonist practices within tourism (Figure 4.3).

The zonkey as an unethical element of heritage

Heritage is a concept that refers to the material, political, economic, and cultural interests of the dominant classes, who cast their stamp of power on the resource of choice (González & Hiernaux, 2015). It has been an instrument to reinforce identity and promote social cohesion (Troitiño, 1998). It is also the expression of artistic, philosophical, and moral values that make for tradition. However, the role of heritage has changed in the last few years due to the prevalence of the economic over other values. Conscious as they were of the abuse and maltreatment to which the donkeys were subjected, the local population rejected the declaration of the zonkey

Figure 4.3: Tourists and the zonkey, 2018.
Source: CharlyPi, 2018. Used with permission.

as cultural heritage, as it was demonstrated in some comments through the online platform of El Universal, one of Mexico's main newspaper:

> They [donkeys] must be freed. They have been exploited long enough and the poor things are exhausted out of standing in the scorching sun, that's sheer cruelty! Let the master take their place and paint himself as a zebra and be mounted by tourists. We'll see if he can take it to stand forever with no food or drink. Ignorant, backward-minded individuals!
>
> (El Universal, 2015)

> Since when is animal abuse a kind of work? Animals are not to take the brunt for their masters' inability to put their lives upright! (El Universal, 2015)

In 2015, an animal rights group petitioned a writ (*amparo*) to reverse the declaration of the zonkey as heritage. The arguments were not only that donkeys have to remain standing for over 12 hours a day under the scorching sun – which clearly affects their well-being – but also that their use as resources for tourists' enjoyment and the treatment they receive goes against the city's laws against animal abuse (Merlo, 2016). As an alternative, this group proposed to use artificial donkeys made of synthetic materials (Notimex, 2015).

Conclusions

The use of non-human animals to attract visitors is a common practice in tourism. The images thus created are not only attractive to tourists, but they also propagate a message about local culture. In Tijuana, for example, representations of the zonkey are found everywhere, in the form of drawings, engravings, photographs, and sculptures across the urban area, and the non-human animal has become a landmark and an identity icon.

The commodification of the zonkey as a tourist resource and heritage element is the product of the unethical supremacist ideology of speciesism, which feeds upon the discrimination against non-human animals entrenched in belief systems, interests, values, and customs. To end this injustice, all legal resources must be put to work at the local, state, and national levels to prohibit these recreational practices.

Commodifying a sentient being is tantamount to turning it into a resource without its consent. The label of heritage element encourages the exploitation to which these non-humans are subjected, especially when the aim is to satisfy tourists' hedonism. From the perspective of an amplified moral consideration of what is right and what is wrong, this is clearly an unethical practice that goes against the basic interests of donkeys and which should be abolished or transformed, for the following reasons: 1) it is unethical that a sentient being be objectified and turned into a commodity for tourists' use; 2) the declaration as "heritage" of a work animal is no justification for its use to obtain economic gain; 3) collective identity should not be built on the abuse and maltreatment of other animals. Although Mexican law considers animal abuse a crime, it continues to be "speciesist", because it fails to protect all animal species from the aggressions they endure every day. In fact, the law favors, promotes, and normalizes the systematic violence against non-human animals when it comes to declaring cultural heritage. For as long as the animals-as-human-property paradigm continues to exist, the Mexican legal system will be a useless tool to ensure the well-being of non-humans.

References

Adams, C., & Gruen, L. (2014). *Ecofeminism: Feminist intersections with other animals and the earth*. New York: Bloomsbury Publishing.

Andrade, J. Moreno, O., & Quiñones, R. (2015). Valoración de atributos culturales e históricos en la imagen promocional de la ciudad de Tijuana como destino turístico. *El Periplo Sustentable: revista de turismo, desarrollo y competitividad*, *28*, 31–58.

Arluke, A. (2002). Animal abuse as dirty play. *Symbolic Interaction*, *25*(4), 405–430.

Belz, F. & Baumbach, W. (2010). Netnography as a method of lead user identification. *Creativity and Innovation Management*, *19*(3), 304–313.

Berumen, H. (2003). *Tijuana la horrible: entre la historia y el mito*. Tijuana: Colegio de la Frontera Norte.
Blanco, S. (2013). ¿Trato o uso? El fin de la explotación animal. *Dilemata, 13*, 259–264.
Bramwell, B., & Meyer, D. (2007). Power and tourism policy relations in transition. *Annals of tourism research, 34*(3), 766–788.
Buller, H. (2015). Animal geographies II: methods. Progress in Human Geography, 39(3), 374–384.
Caballero, Y. (2014). Declaran al "burro-cebra" patrimonio de BC. *La Crónica* [online]. Retrieved September 17, 2019, from https://www.lacronica.com/EdicionEnLinea/Notas/Noticias/11082014/873610-Declaran-al-burro-cebra-patrimonio-de-BC.html
CarlyPi (2018). Tijuana 2018. *Twitter*, [online]. Retrieved September 17, 2019, from https://twitter.com/CharlyPi/status/1046157158188867584
Carmona, R. (2003). *Organización territorial del turismo en Tijuana, Baja California*. Mexico City: Facultad de Filosofía y Letras, Universidad Nacional Autónoma de México.
Carmona, R., & Correa, O. (2008). Estructura territorial del turismo en el corredor Tijuana-Rosarito-Ensenada. *Teoría y Praxis, 5*, 359–375.
Carrillo, J. & Hualde, A. (2000). Desarrollo regional y maquiladora fronteriza: peculiaridades de un clúster electrónico en Tijuana. *El mercado de valores, 60*(10), 45–56.
Ceballos, M. (2001). *Encuentro en la frontera*. Tijuana: El Colegio de México.
Cháidez, J. (2012). Las Cebras de Tijuana. *Leyendas de Tijuana y sus alrededores: mitos, y otras cosas que platica la gente*. [online]. Retrieved September 22, 2019, from http://leyendasdetijuana.blogspot.com/2012/11/las-cebras-de-tijuana.html#ixzz4MoeZBovY
Cheong, S. & Miller, M. (2000). Power and tourism: A Foucauldian observation. *Annals of Tourism Research, 27*(2), 371–390.
Christ, C. (2006). Ecofeminism and process philosophy. *Feminist Theology, 14*(3), 289–310.
Cohen, E. (2005). Principales tendencias en el turismo contemporáneo. *Política y sociedad, 42*(1), 11–24.
Cottingham, J. (1978). 'A Brute to the Brutes?': Descartes' Treatment of Animals. *Philosophy, 53*(206), 551–559.
COTUCO (2012). Comité de Turismo y Convenciones de Tijuana 2012. *Portal oficial de promoción de la ciudad de Tijuana* [online]. Retrieved September 17, 2019, from http://www.descubretijuana.com/
Culture Trip (2016). A young boy posing with the Tijuana "zonkey" in 1949. *Culture Trip* [online]. Retrieved September 17, 2019, from https://theculturetrip.com/wp-content/uploads/2018/03/98879613_3dacb1691f_o.jpg.
Dawkins, M. S. (1990). From an animal's point of view: Motivation, fitness, and animal welfare. *Behavioral and Brain Sciences, 13*(1), 1–60.
Decrop, A. (1999). Triangulation in qualitative tourism research. Tourism management, 20(1), 157–161.
El Universal (2015). ONG busca desaparecer en Tijuana a "burros-cebra". *El Universal* [online]. Retrieved September 22, 2019, from https://www.eluniversal.com.mx/articulo/estados/2015/09/12/ong-busca-desaparecer-en-tijuana-burros-cebra.
Fennell, D. (2012). Tourism, animals and utilitarianism. *Tourism Recreation Research, 37*(3), 239–249.
Fennel, D. & Malloy, D. (2007). *Codes of ethics in tourism: Practice, theory, synthesis*. New York, Channel View Publications.
Fennell, D. & Nowaczek, A. (2010). Moral and empirical dimensions of human-animal interactions in ecotourism: deepening an otherwise shallow pool of debate. *Journal of Ecotourism, 9*(3), 239–255.

Francione, G. L. (2009). *Animals as persons: Essays on the abolition of animal exploitation.* New York: Columbia University Press.

Francione, G. L., & Garner, R. (2010). *The Animal Rights Debate: Abolition or Regulation?* New York: Columbia University Press.

González, C. & Hiernaux, D. (2015). Patrimonio y centralismo: perspectivas críticas desde el ámbito local. *Revista A*, *190*, 1824–2014.

Harari, Y. N. (2016). *Homo Deus: breve historia del mañana*. Mexico City: Debate.

Horta, O. (2017). *Un paso adelante en defensa de los animales*. Madrid: Plaza y Valdés.

Hribal, J. (2003). Animals are part of the working class: a challenge to labor history. *Labor history*, *44*(4), 435–453.

Ianni, O. (1996). *Teorías de la globalización*. Mexico City: Siglo XXI.

INEGI (2018). Baja California: División Municipal. *INEGI Cuéntame* [online]. Retrieved July 3rd, 2020, from http://cuentame.inegi.org.mx/monografias/informacion/bc/territorio/div_municipal.aspx?tema=me&e=02

LAPIZTOLA (2010). A bañar al burro zebra! *Blogspot* [online]. Retrieved September 17, 2019, from http://lapiztola.blogspot.com/2010/12/banar-al-burro-zebra.html

López, A. & Bringas, N. (2016). Turismo. Una visión desde el grado de concentración territorial municipal de la actividad turística. In J. O. Moncada & A. López (Eds.). *Geografía de México. Una reflexión espacial contemporánea* (pp. 730–745). Mexico City: Instituto de Geografía, UNAM.

Lynn, W. (1998). Animals, ethics and geography. In J. Wolch & Emel, J (Eds.) *Animal geographies: Place, politics, and identity in the nature-culture borderlands* (pp. 280–297). New York: Verso Books.

Markwell, K. (2015). *Animals and tourism: Understanding diverse relationships*. London: Channel View Publications.

Martínez, L., & Contreras, D. (2017). La tradición del llamado "burro cebra" genera polémica en Tijuana. *Telemundo 20* [online]. Retrieved September 18, 2019, from https://www.telemundo20.com/boletin-electronico/noticias/La-tradicion-del-llamado-burro-cebra-genera-polemica-en-Tijuana-445549543.html

Mendl, M. & Paul, E. (2004). Consciousness, emotion and animal welfare: insights from cognitive science. *Animal Welfare*, *13*(1), 17–25.

Merlo, E. (2016). En abril dirán si burro-cebra sigue como patrimonio cultural de BC. *UNIRADIO* [online]. Retrieved September 21, 2019, from https://www.uniradioinforma.com/noticias/bajacalifornia/390724/en-abril-diran-si-burro-cebra-sigue-como-patrimonio-cultural-de-bc.htm

Newsweek Mexico (2017). El "burro-cebra" podría volver a ser patrimonio cultural de BC. *Newsweek Mexico* [online]. Retrieved September 18, 2019, from https://newsweekespanol.com/2017/09/burro-cebra-podria-volver-a-ser-patrimonio-cultural-de-bc/.

NOTIMEX (2014). Tradición de "burro-cebras" en Tijuana cumple cien años. *Publimetro* [online]. Retrieved September 22, 2019, from https://www.publimetro.com.mx/mx/vida/2014/08/02/fotos-tradicion-burro-cebras-tijuana-cumple-cien-anos.html.

NOTIMEX (2015). ONG busca desaparecer en Tijuana a "burros-cebra". *El Universal* [online]. Retrieved September 22, 2019, from https://www.eluniversal.com.mx/articulo/estados/2015/09/12/ong-busca-desaparecer-en-tijuana-burros-cebra.

Observatorio Turístico de Baja California (2015). *El burro-cebra de Tijuana: Testigo mudo de la evolución creativa de la ciudad*. Mexicali: Secretaría de Turismo del Estado de Baja California.

Organización Mundial del Turismo (1999). *Código Ético Mundial para el Turismo*. New York: Organización Mundial del Turismo.

Palaversich, D. (2002). La ciudad que recorro. Un flâneur en Tijuana. Literatura Mexicana, 13(2), 215–227.

Quintero, G. J. & López, A. (2018). Tauromaquia y turismo oscuro en México: las corridas de toros como prácticas no éticas. *Teoría y Práxis 24*. 197–228.
Rebollo, I., Polderman, T., & Moya, L. (2010). Genética de la violencia humana. *Revista de neurología, 50*(9), 533–540.
Regan, T. (1987). The case for animal rights. *Advances in Animal Welfare Science*, 1986/87, 179–189.
Reyes, K. (2017). Mantendrán lucha por burro-cebra como patrimonio. *Frontera, México* [online]. Retrieved September 17, 2019, from https://www.frontera.info/EdicionEnlinea/Notas/Noticias/19092017/1257068-Mantendran-lucha-por-burro-cebra-como-patrimonio.html.
Ruiz, J., Martínez, O., & Verján, R. (2015). Valoración de atributos culturales e históricos en la imagen promocional de la ciudad de Tijuana como destino turístico. *El Periplo Sustentable: revista de turismo, desarrollo y competitividad, 28*(2), 31–58.
Singer, P. (1975), Animal Liberation, Avon: New York, NY.
Singer, P. (2009). Speciesism and moral status. *Metaphilosophy, 40*(3, 4), 567–581.
Skolimowski, H. (2017). *Filosofía Viva. La ecofilosofía como un árbol de vida.* Girona: Atalanta.
Stolk, H. (2004). 'This Is Not Mexico, This Is The Border': Discourses on Authentic Mexican Culture in Tijuana. *Etnofoor*, 227–242.
Swann, W. J. (2006). Improving the welfare of working equine animals in developing countries. *Applied Animal Behaviour Science*, 100 (1–2), 148–151.
Taylor, N., & Signal, T. (2009): «Pet, pest, profit: Isolating differences in attitudes towards the treatment of animals», Anthrozoös, 22(2), 129–135.
Tovar, L. (2015). Consentimiento. *Filosofía Vegana* [online]. Retrieved September 30, 2019, from http://filosofiavegana.blogspot.com/2015/10/consentimiento.html.
Troitiño, M. (1998). Turismo y desarrollo sostenible en ciudades históricas. *Ería: Revista cuatrimestral de geografía, 47*(1), 211–228.
Uniradio Informa (2017). Controversia por figura del Burro-Cebra como patrimonio cultural en BC. Uniradio [online]. Retrieved September 17, 2019, from https://www.uniradioinforma.com/noticias/reportajesespeciales/494720/controversia-por-figura-del-burrocebra-como-patrimonio-cultural-en-bc.html.
Upchurch, R. (2000). Código de ética. *Enciclopedia de Turismo*. Madrid: Síntesis.
Urbanik, J. (2012). *Placing animals: An introduction to the geography of human-animal relations*. New York: Rowman & Littlefield.
Urbanik, J. & Johnston, C. (2017). *Humans and Animals: A Geography of Coexistence*. New York: ABC-CLIO.
Vaquera, S. (2007). Postcards from the Border: In Tijuana, Revolución is an Avenue. In Manzanas, A. (Ed.) *Border Transits: Literature and Culture Across the Line* (pp. 79–97). Amsterdam: Brill-Rodopi.
Vincent, J. (2018). How Zebra-painted Donkeys Became Tijuana's Hottest Tourist Attraction. *Culture Trip* [online]. Retrieved September 17, 2019, from https://theculturetrip.com/north-america/mexico/articles/how-zebra-painted-donkeys-became-tijuanas-hottest-tourist-attraction/.
Walker, M. (2011). Knowledge production and border nationalism in northern Mexico. *Nations and Nationalism, 17*(1), 168–187.
Walton, J. (1997). Review articles: Taking the history of tourism seriously. *European History Quarterly, 27*(4), 563–571.
Webster, A. (2001). Farm Animal Welfare: the Five Freedoms and the Free Market. *The veterinary journal, 161*(3), 229–237.
Weitzenfeld, A. & Joy, M. (2014). An overview of anthropocentrism, humanism, and speciesism in critical animal theory. *Counterpoints*, 448, 3–27.

Wolfe, C. (2010). *What is posthumanism?* Minneapolis: University of Minnesota Press.
Xiao, L., & Zhao, L. M. (2009). The tourism destination image of Taiwan disseminated on Internet: Based on a content analysis of travel related websites across Taiwan straits. Tourism Tribune, 3(24), 75–81.
Zenteno, R. (1995). Del rancho de la Tía Juana a Tijuana: una breve historia de desarrollo y población en la frontera norte de México. *Estudios demográficos y urbanos*, *3*(2), 105–132.

Álvaro López-López, Gino Jafet Quintero Venegas

5 Animal dark tourism in Mexico: Bulls performing their own slaughter

Introduction

Speciesism, the notion that all other species are inferior to humans, is no different from other irrational discriminatory behaviors, such as racism or sexism, except for the fact that humans are also animals (Harper, 2011; Kappeler, 1995; Singer, 2009; Steinbock, 1978). Adhering to a post-humanist ethic, in this chapter the expressions "non-human animals" and "human animals" will be used to emphasize the fact that members of the animal kingdom, including bulls, are sentient beings and self-conscious of their own pain and pleasure (Low, 2012).

In Mexico, bullfighting as a tourist attraction (*tauroturismo,* in Spanish) incorporates various activities, such as visiting bullrings or thematic museums, registering for workshops, buying souvenirs, and attending the bullfight performance. The latter is the most profitable activity of all–and, unless otherwise stated, the one we will refer to when we use the term "bullfighting." According to Bolio (2017) and López (2020), the Mexican Ministry of Agriculture and Rural Development (Sagarpa) registered 504 bullfights in 2016 in the country, where 2,725 bulls were killed. The annual revenue was 5,541,040,491 Mexican pesos (equivalent to USD $378,952,606).

According to Urbanik (2012), the bulls that are sentenced to death in the arena are working animals of the "entertainment" subcategory that, unlike other animals – whose life is usually prolonged, along with their exploitation – are rather valued for their death in a macabre performance. Further, the bulls demonstrate Coulter's (2016) "dirty work" which is low status, demeaning work, as well as "body work" which involves physical labor. Thus, given its focus on violence and murder for the delight of tourists, the authors of this chapter propose that bullfighting be framed as dark tourism. For this, some barriers must be overcome. Dark tourism is related to the supply or demand of spaces where death is present in a real or represented way (Lennon & Foley, 2000; Stone & Sharpley, 2008). Although for Sharpley (2009), ethics is one of the four most relevant topics of the analysis of dark tourism, however his approach is anthropocentric.

The occasional reference to non-human animals in dark tourism is always as circumstantial objects accompanying human deaths. So, the question arises: why is the death of non-human animals absent in the ethical reflections of dark tourism? Largely because these studies have been framed in scientific humanism, the dominant

Acknowledgment: The cartography included in this chapter was elaborated by Lourdes Godínez Calderón.

https://doi.org/10.1515/9783110664058-005

framework for scientific production in the social sciences, which regards animals as resources or merchandise, outside any ethical consideration (Armstrong, 2008; Best, 2009; Franklin & White, 2001; Wolfe, 2009).

Ethical reflection on bullfighting in dark tourism calls for appreciation of the bulls' existence from a post-humanist ethical stand (Cudworth, Hobden, & Kavalski, 2018). Post-humanist ethics maintains that human and non-human animals should be regarded as equals (Low, 2012), and that in order to make the latter subjects of moral interest, the idea of ethical superiority of humans in relation to other animals must be eliminated (Faria, 2016). The unethicalness of bullfighting is thus revealed: since humans are not above the primary interests of animals, in general, putting bulls to perform their own death for the amusement of tourists is an aberration (Regan, 1983).

The central objective of this chapter is to expose the unethical nature of bullfighting. Following documentary research, which helped to understand the links between dark tourism, bullfighting, and the ethical consideration of death as work, we have explored tourists' perceptions of the performance in Mexico focusing on the spectators' take on cultural heritage, masculinity, and ethics. The information was obtained from: 1) a semi-closed survey conducted in the following Facebook pages: "Aficionados Taurinos del Mundo" ("Bullfighting Fans of the World") (2017), "México Taurino" ("Bullfighting Mexico") (2017), "Toros y Toreros" ("Bulls and Bullfighters") (2017), and "Taurinos Unidos" ("Bullfighters United") (2017); 2) the virtual discourse (Leong, Joseph, & Boulay, 2010) from the comments on the following TripAdvisor pages: "Plaza de Toros Cancún" ("Bullring Cancun") (2019a), "Monumental Plaza de Toros México" ("Monumental Bullring Mexico") (2019b), and "Plaza México" (2019c). The comments were classified according to the gender and cultural circumstances of the commentators (Anderson, Adey, & Bevan, 2010), and the chosen ones were paraphrased here.

The following section grapples with the relevance of bullfighting as a topic of interest for dark tourism from a post-humanist perspective. Next we explore the work of fighting bulls and their job to delight tourists with their own death in the arena. Then, we analyze bullfight tourists' perceptions and discourses, focusing on cultural heritage, masculinity, and ethical issues. Finally, in the conclusions, the unethical nature of the work that has been forced upon bulls – one that leads to their own slaughter – is exposed from a post-humanist ethics perspective that values sentient and self-conscious life.

Dark tourism, bullfighting, and post-humanist ethics

Since the 1990s, when the concept of dark tourism first appeared in world literature, several coinciding definitions have emerged. Foley and Lennon (1996) defined dark tourism as the consumption of sites of death and disasters, real or represented, while Stone (2006) referred to it as a deliberate or circumstantial trip to macabre sites of death and disaster. Based on recent conceptual discussions, Van Broeck and López (2018) concluded that dark tourism is characterized by the commodification of death – real or represented.

Studies on dark tourism encompass war museums, cemeteries, and memorials (Baldwin & Sharpley, 2009); concentration camps (Miles, 2002); sites of destruction and major disasters (Pezzullo, 2009); and places where murders or accidents of famous people have taken place (Foley & Lennon, 1996). What these studies have in common is the anthropocentric focus on human death, evidencing the predominance of humanism in science (Armstrong, 2008; Best, 2009; Franklin & White, 2001; Wolfe, 2009). In humanism, culture is positioned as a supreme value, which is achieved to the extent that human beings move away from their animality (Elias, 2016). Thus, non-human animals are seen as inferior and, in this way, their commodification is justified (Horta, 2017; Hribal, 2003).

On the ethical issue of the human-non-human divide, the works of Singer (1975) and Regan (1983) contributed to change the focus of the discussion to sentience, a criterion to understand that avoiding pain and suffering, and preserving life, were as much the interests of humans as of non-humans. Later on, the Cambridge Declaration on Consciousness (Low, 2012) has provided scientific demonstration that non-human animals possess a central nervous system that equips them with the awareness of pain and pleasure. This has reinforced a post-humanist ethic tending to match humans to their animal otherness (Faria 2016; Urbanik & Johnston, 2017).

Post-humanist ethics have revealed the power relations that human beings have imposed over non-humans through their commodification and enslavement, including their use as tourist attractions (Bertella, 2013; Fennell, 2012; Markwell, 2015). As for dark tourism, Light (2017) calls for new academic perspectives, given that the analysis of death in tourism has so far been rather limited and confined to narrow cultural and geographical parameters. The slaughter of a bull in bullfighting as the center of tourists' attention is a legitimate topic for analysis within the framework of dark tourism focusing on ethical issues.

Bullfights are a recognized component of Iberian culture that has spread to other parts of the world, notably Latin America. The discourse that sustains this practice is that "brave" men face equally "brave" bulls in a closed enclosure (*ruedo*) under a protocol that is presumed to be aesthetic (Cohen, 2014; Douglas, 1999; Landborn, 2015; Mitchell, 1991). The modern version of bullfighting dates from late 18th century Spain, when standards were set so as to offer a performance full of tricks and maneuvers that ensure the death of bulls and not bullfighters (Douglas, 1999).

Bullfighting is commercialized as "blood sport," "bullfighting art," or "wild party" ("*fiesta brava*") (Douglas, 1999; Mitchell, 1991; Rangel, 2004), and in tourism bullfights are promoted as exotic otherness (Brandes, 2009; Mestre, Del Rey, & Stanishevski, 2008). Although tourism and bullfighting are best articulated in

Spain (Tauroturismo, 2017), in Mexico there is a very important tourist dynamic around bullfighting that combines festiveness with death.

There are 15 cities with relevant bullrings in Mexico which, according to Quintero and López (2018), have been mentioned by travelers as positive or negative tourist attractions (Figure 5.1). Some of them belong in cities of strong Spanish heritage, like Aguascalientes, Tlaxcala, Morelia, or Merida, where bullfighting performances are not only popular among locals, but very important tourist attractions during famous festivities, such as the National Fair of San Marcos, the Tlaxcala Fair or the Yucatan Fair; while others are located in cities where bullfighting is only one in a wide range of tourist attractions, like Mexico City and Cancun. Because these two cities receive the largest number of tourists in the country, including bullfighting tourists (*tauroturistas*), they offer not only large bullrings but also complementary infrastructure, such as museums and restaurants catering to this kind of tourism (Quintero & López, 2018). This is examined later in the chapter through analysis of the information provided by tourists who have attended bullfights in these two urban areas.

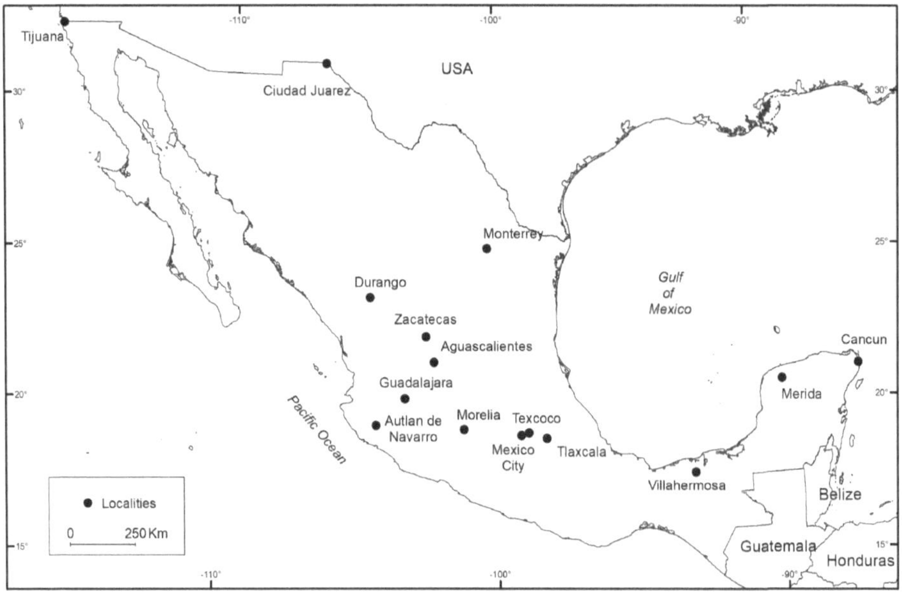

Figure 5.1: Mexican cities with bullrings of tourism importance in 2018.
Source: Lourdes Godinez. Used with permission.

Bulls as workers in the bloody performance of bullfighting

Animals working in tourism are often forced to perform repetitive, visible, and normalized tasks in a certain space for most of their lives (Urbanik, 2012), but in bullfighting bulls' work is ephemeral, and their job is to excite spectators through the process of their own bloody death. The assignment of a bull to a bullfight depends on a long selection process aimed at choosing the more "aggressive" exemplars, although the truth is that their natural behavior has nothing of the supposed aggression attributed to them (Ortiz, 2014). Vanda, Edwards, and Cossío (2017, p. 4) present a detailed forensic report carried out on bulls having died during bullfights, and based on Purroy and González (1985) and Muñoz, Agüera, and Castejón (2007), these authors conclude that the extreme pain resulting from the physical attack on these non-human animals produces serious mental damage, including fear, anxiety, anger, and frustration. As in any other mammal, this process is activated by the release of neurotransmitters and hormones such as catecholamines (adrenaline and norepinephrine) and glucocorticoids (such as cortisol).

The selection of bulls for fighting is a process that takes four years. Upon reaching their first year of age, calves are weaned, tagged, and confined in single sex groups (Saborit, 2010). One year later, they are sent to a trial (*la tienta*), in which only those who are brave enough to withstand the pain inflicted by a picador are allowed to go on living, the others being sent to the slaughterhouse (Saborit, 2010; Medina, 2018). When they reach their third year of age, they are called steers (*novillos*), and they may be destined either to be slaughtered by bullfighters during private training or wait for another year before being sent to a bullfight in public (ABC. ES, 2018; Kalof, 2014; Saborit, 2010).

For a bull, entering the bullring is like a death sentence, and no more than 20 minutes will pass before he is slaughtered. However, a *corrida* (bullfighting "spectacle") consists of the death of six bulls, two per bullfighter (Cohen, 2014; Elizalde, 2012). Present in the ring are killer bullfighters, junior bullfighters, *alguacilillos* (sheriffs), picadors, *mulilleros* (draggers) and *monosabios* (helpers), a total of about 25 humans, and about 16 non-human animals, including bulls, horses, and mules (Servitoro, 2019). Each episode of a *corrida*, in which a bull is killed, is called *faena* (labor, slaughter), which in turn is divided into three *suertes* (dexterities) or acts announced by the sound of a trumpet.

During the *tercia de varas* (third of canes), the bullfighters tire the bull with a cape and then from atop their horses the picadors throw goads at him with sharp tips to cut his muscles and weaken him. In the "third of flags" the bull is wounded with sticks of about 70 centimeters that end in metal blades to dramatically hurt his internal organs. In the "third of death," when the bull is already bleeding, sore, and with many vital problems, a sword is stuck between his shoulder blades, but the agony of his death usually lasts many minutes, and even when the excited audience asks that the tail and ears be cut to make a gift to the bullfighter, it is common that the bull is still conscious (Elizalde, 2012; Ortiz, 2014; Vanda, Edwards, & Cossío, 2017).

During the *corrida*, the bull's job is to demonstrate his "strength" and "bravery," which is contrasted with the masculinity of a bullfighter who is always appreciated as superior, not only because of his supposed intelligence but because he can bend and slaughter a "wild" bull in an "artistic" way. Theoretically, the bulls should go out full of zest into the ring in a natural way, to show their good breeding (Kalof, 2014), but it is known that before entering the ring they are injured and manipulated clandestinely, and that during the run they suffer inexpressible physical and psychological damage (Vanda, Edwards, & Cossío, 2017). This massacre performance is presented as an event with sophisticated heritage, and the characters involved, their costumes, calculated movements, music, etc., affect the perceptions that tourists have of this practice, some of them positive, some of them negative, as will be seen in the following section.

Tourists' perceptions of a corrida

The cultural discourse

The central and defining element of dark tourism is the commodification of death as a cultural expression, and the ethics of this is one of the main topics of interest for researchers (Sharpley, 2009). The endorsement or rejection of bullfighting as a cultural practice in tourism depends on the philosophical paradigm in which the analysis is laid to rest: humanism versus post-humanist ethic. These two positions will be used as reference to analyze the virtual discourse of those who have attended a *corrida* and commented on it in the aforementioned Facebook and TripAdvisor pages.

As can be seen in the following comments, virtually no reference is made of the bulls' suffering in the bullfighting spectacle, and the prevailing discourse is one of bullfighting as a "sophisticated" cultural tradition or an exciting experience:

> Tourist 1 (male, Mexico) said that every year he attends bullfights at the Plaza de Toros Mexico, and he feels very proud about it because being the biggest bullfighting ring in the world, visiting it fills him up with culture; Tourist 2 (male, Mexico) said he travels through Mexico in order to attend bullfights, which he regards as the most cultured and poetic event in the country; Tourist 3 (male, Mexico) declared that he considers himself an intellectual because he attends bullfights, and stated that if others don't understand the event, it is out of ignorance; finally, Tourist 4 (female, Mexico) said that she attends bullfights because of passion and emotion, and not for recreational purposes.

Artists and scholars of Spanish tradition in both Spain and Latin America have, intentionally or not, contributed to legitimizing and promoting bullfights, which entrenches this practice within the tourism market. Painters like Pablo Picasso, Francisco de Goya, Salvador Dalí, and Fernando Botero; writers including Ernest Hemingway and Gabriel García Márquez; filmmakers such as Carlos Saura and Pedro Almodóvar; and philosopher Fernando Savater, among many others, have contributed to build the narrative of bullfighting as a form of art (Mosterín, 2010; Salazar, 2013; Savater, 2010).

The relevance of these artists and humanists has been used the world over as *ad verecundiam* ("if the master says it, it must be true") and *ad antiquitatem* ("if something has been done since time immemorial, it must be right") arguments by virtually all the actors in the tourism business (promoters, business people, tourists, local inhabitants) to justify the practice of bullfighting as part of the cultural patrimony, to the point of making it incontrovertible and incontestable (Markwell, 2015; Ortiz, 2014; Wilson, 1998). This is the bias of the following comments:

> Tourist 5 (male, Mexico) stated that a bullfighting ring is a charming traditional place where one delights at the colors and the hues of the bullfighting art; Tourist 6 (male, United States), said that bullfights have been a cultural event for many decades and therefore tourists must be open-minded if they attend; Tourist 7 (male, United States) pointed out, in a very similar fashion, that attending a bullfight wasn't a problem at all, as he links it to a long-standing local culture.

Although bullfighting is regarded as alien outside Spain, Portugal, and Latin America, Lewine (2005) points out that tourism has managed to exploit it as an attraction for foreign visitors that hinges on a narrative of ritual, death, and art. Yates (2009) reports the yearly travel of US men to Spain and Latin America to attend bullfights, inspired by Hemingway's writings. In several opinions by foreign tourists from outside Latin America recorded in this survey, a certain awareness of the cruelty to which bulls are subjected was detected along the bullfighting-as-culture/art discourse.

This awareness, however, seems not to be strong enough to act as a deterrent. This seems to be the case with the following travelers who, despite being conscious of the fact that bullfighting is a blood-exhibition affair leading to the bull's slaughter, try to justify it:

> Tourist 8 (female, United States) affirmed that human rights activists should not attend bullfights, and that in order to enjoy the show, one must accept this tradition; Tourist 9 (male United States), though recognizing the cruelty involved in the slaughter of a bull – and meanwhile making fun of his wife's aversion to bullfights – argued for the appreciation of this Mexican tradition and the "cool experience" it provides; Tourist 10 (female, United States) argued that those who think of bullfighting as animal cruelty should not attend, but those who see it as tradition and as a cultural element, will appreciate the beauty of the delicate movements of the bullfighters in their suits of lights, and ended by adding that she brought her 17-year-old daughter to the show, who now has become a fan.

The discourse of masculinity

The socially constructed idea of masculinity is based on obstinacy, assertiveness, sexual potency, independence, and self-control (Stafford, 1998). In bullfights, the symbolic values of manliness are reinforced when a matador is capable of confronting and dominating a bull (Kalof, 2014; Ortiz, 2014; Thompson, 2013). By the act of slaughtering, the torero is elevated to a position of superiority, with the endorsement of the majority of spectators (male and female), while the role of the bulls, as part of the non-human working class, is to satisfy this need to reinforce the construction of masculinity. Cohen (2014) has pointed out that many tourists reinforce their masculinity by engaging in hunting and attending bullfights. During a trip, men reinforce their masculinity by, among other things, interacting with risk, adventure, and extraordinary experiences (Lozanski, 2015; Thompson & Holt, 2004). The empathy of most tourists lies with the torero, who incarnates the winner, a symbol of masculinity.

Ecofeminism maintains that the unequal relationships between men and women are transferred to the existing power relationships between "human beings" (who incarnate the masculine) and "nature" (who incarnates the feminine), and that in this relationship, favorable to the masculine, humans take on themselves the right to dispose of animal otherness (Adams & Gruen, 2014; Puleo, 2014; Salleh, 1997). This is evident in the comment by Tourist 11 (male, Mexico), who justifies the "right" of the *torero* to kill a bull because he is exposed to the danger of dying in front it – which is a fallacy because, as it has already been said, the chances of a *torero* dying are virtually zero. He ended by adding that a bullfighter is the incarnation of a true man, who is capable of risking his own life.

Visitors regard the *torero* as a hero, whose masculinity has been proven in the ring by facing what is considered a dangerous beast – however, unlike lions, for example, this bovine is not even carnivorous: "bullfights have a glorifying image: they are presented as a competition between the brave matador, who valiantly risks his life and physical integrity to confront a furious beast" (Difusión Animalia, 2015, s/p). Some tourists attribute the *torero*'s hyper-masculinity to his self-control of the fear of imminent danger and death, as did Tourist 12 (male, Mexico) when saying that bullfight rings are the ultimate place on Earth where man demonstrates his balance with nature, as he manages to face the threat posed by the bull, the multiple scars that he (the *torero*) sports on his body being a proof of it.

Clad in his awesome, groin-enhancing *traje de luces* (lights attire), the matador's performance depends not only on his ability to face danger, but also on his intelligence to cheat the bull. The more difficult it is to dominate the bull, the more daring the matador's performance becomes, and the more obvious his bravery is (Douglas, 1999; Thompson, 2013).

The ethics of rejecting bullfights

In the context of an increasing awareness of animal abuse in tourism and the measures to prevent it (Özlem y Çiftçi, 2013), it is surprising that tourists continue to endorse the violence of bullfighting – though some do recognize the ethical problems involved – by simply alluding to cultural patrimony arguments. The prevalence of a dominant anthropocentric positioning seems to be the rule. However, since some tourists may have attended a *corrida* without really knowing what to expect, there are also positions against the practice.

> Tourist 13 (female, United Kingdom) declared that she cannot erase from her mind the terrible image of the bull collapsing and convulsing, blood dripping from his body; Tourist 14 (male, United States) considered that the bullfighter's performance has nothing to do with bravery, and that the same applies to hunters and animal abusers; Tourist 15 (male, Mexico) affirmed that this show is at the same level of cruelty as the one prevailing at the times of the Roman Empire; and Tourist 16 (male, United Kingdom) regretted to have been so naïve as to not knowing that during bullfights, the bulls are tortured until they are killed, and confided to have gone into a shock when he realized it. He ended up by celebrating that this barbarity has been outlawed in most countries.

Following Ortiz (2014), Savater (2010) justifies in *Tauroética* the practice of bullfighting with these arguments: a) humans have no moral obligations to the bull, a non-rational being; b) bullfighting may be problematic not because of the bull itself, but because it could develop cruelty towards human beings; c) the life of bulls is not morally relevant; d) if bullfighting were to be abolished, the cultural landscapes and the traditions linked to it would disappear; e) the anti-bullfights position is very arrogant, for it seeks to impose its abolition with no regard for the right of fans to enjoy this show.

But Ortiz (2014) counteracts Savater on ethical grounds: "I maintain that he is mistaken and that his moral arguments lack the strength he attributes to them; there may be economic, political, and other reasons to keep *corridas* going, but certainly not moral reasons" (p. 205). This author goes on to point out that in a private survey, 80% of the Mexican population is against bullfights, but that the weight of the economic interests involved and the influence of several groups of "hispanophiles," notably entrepreneurs and politicians, has prevented its abolition (Chávez, 1999; Elizalde, 2012; Ortiz, 2014). The unethicalness of this type of work imposed on the bulls is so blatant, that it must be abolished right away.

Conclusions

The ethical dimension of bullfighting as a tourist attraction has yet to be submitted to careful analysis through the lens of dark tourism. From a post-humanistic ethical perspective, it is clear that the tourists that have attended and supported bullfights have built their arguments on the discourse of masculinity and cultural patrimony. The use of the *ad verecundiam* and the *ad antiquitatem* fallacies is common in their comments on social media, which legitimate this variety of zoo-slavery.

The works by philosophers like Regan (1983) and Singer (1975), and the evidence provided by neurologists like Low (2012) and others, have allowed to firmly establish the sentience and self-consciousness of non-human animals, which makes any kind of zoo-slavery ethically untenable, and all the more so when the end is to inflict deep suffering and death on them. Though from an ontological viewpoint pain is a natural occurrence in both human and non-human animals, to deliberately inflict pain on a sentient being with the purpose of satisfying a ludic interest, is ethically unsound, no matter if this practice has become a tradition. As Ortiz (2014) points out, if we have the cognizance to recognize a "good" tradition for human beings, it is because "bad" ones also exist, and we must eradicate them.

Slaughtering bulls is not only an anthropocentric action, but a sexist one too, for it exalts masculine attributes such as aggressiveness, dominance, and contempt for nature and alternative expressions. Bullfights reinforce patriarchal thinking, which, from an ecofeministic perspective, has been responsible not only for the exploitation of women, but also for the exploitation of nature and the treatment of non-human animals as resources, leading to zoo-slavery. The notion of manliness was present in most of the comments on social media by tourists who attended bullfights in Mexico, and who were enthralled by witnessing the prototypical image of "a man."

The need emerges to further the study of the use of non-human animals in tourism and prevent their abuse and death, especially when such practices are entrenched into the discourse of heritage. Without any doubt, the primary interests of non-human animals (self-preservation, avoidance of pain) are above the secondary interest of tourists in their own recreation and entertainment. Article 2 of the UNWTO Code of Ethics (1999) should be modified to read thus: "the exploitation of beings [sentient and self-conscious of their pain and pleasure] in any form [. . .] violates the fundamental objectives of tourism and constitutes a denial of its essence." Economic benefit can in no way justify the persistence of slavery, human or non-human.

References

ABC.ES (2018). Los seis requisitos que deben reunir los toros para lidiarse [online]. Retrieved November 2, 2019, from https://www.abc.es/cultura/toros/abci-seis-requisitos-deben-reunir-toros-para-lidiarse-201810241849_noticia.html.

Adams, C., & L. Gruen (2014). *Ecofeminism: Feminist Intersections with Other Animals and the Earth*. New York: Bloomsbury Academic.

Aficionados Taurinos del Mundo (2017). Facebook group site. Retrieved October 14, 2017, from https://www.facebook.com/Aficionados.Taurinos.del.Mundo.

Anderson, J., Adey, P., & Bevan, P. (2010). Positioning place: polylogic approaches to research methodology. *Qualitative Research*, *10*(5), 589–604.
Armstrong, P. (2008). *What animals mean in the fiction of modernity*. London: Routledge.
Baldwin, F. & Sharpley, R. (2009) Battlefield Tourism: Bringing Organized Violence Back to Life. In R. Sharpley & P. Stone (Eds.), *The Darker Side of Travel. The Theory and Practice of Dark Tourism* (pp. 186–206). Bristol: Channel view publications.
Bertella, G. (2013). Ethical content of pictures of animals in tourism promotion. *Tourism Recreation Research*, *38*(3), 281–294.
Best, S. (2009). The rise of critical animal studies: Putting theory into action and animal liberation into higher education. *Journal for Critical Animal Studies*, *7*(1), 9–52.
Bolio, A. (2017) "Caracterización y dimensionamiento del sector bovino, espectáculo en México". *Toriles.com* [online]. Retrieved May 8, 2020, from https://toriles.com/nota%20dtl.php?recordID=165
Bourdieu, P. (1980). *Structure, habitus, practices. The logic of practice*. Stanford: Stanford University Press.
Brandes, S. (2009). Torophiles and torophobes: The politics of bulls and bullfighting in contemporary Spain. *Anthropological Quarterly*, *82*(3), 779–794.
Chávez, M. (1999). *Sociología de la tauromaquia: análisis sociológico de la fiesta brava en México*. Facultad de Ciencias Políticas y Sociales: UNAM.
Cohen, E. (2014). Bullfighting and Tourism. *Tourism Analysis*, *19*(5), 545–556.
Coulter, K. (2016). *Animals, work, and the promise of interspecies solidarity*. Palgrave Macmillan.
Cudworth, E., Hobden, S., & Kavalski, E. (2018). *Posthuman Dialogues in International Relations*. Abingdon: Routledge.
Difusión Animalia (2015). Tauromaquía y Jineteadas [online]. Retrieved November 11, 2017, from http://difusionanimalia.blogspot.mx/2015/05/tauromaquia-y-jineteadas.html
Douglas, C. (1999). *Bulls, bullfighting, and Spanish identities*. Phoenix: University of Arizona Press.
Elias, N. (2016). *El proceso de la civilización: investigaciones sociogenéticas y psicogenéticas*. Mexico City: Fondo de Cultura Económica.
Elizalde, J. (2012). *El arte en la muerte y el sufrimiento: tauromaquia y sacrificio*. Facultad de Estudios Superiores Aragón: UNAM.
Faria, C. (2016). *Animal ethics goes wild: the problem of wild animal suffering and intervention in nature*. Barcelona: Universitat Pompeu Fabra.
Fennell, D. (2012). *Tourism and Animal Ethics: Contemporary Geographies of Leisure, Tourism and Mobility*. Abingdon: Routledge.
Foley, M. & Lennon, J. (1996). JFK and dark tourism: A fascination with assassination. *International Journal of Heritage Studies*, *2*(4), 198–211.
Franklin, A., & White, R. (2001). Animals and modernity: changing human–animal relations, 1949–98. *Journal of sociology*, *37*(3), 219–238.
Hardouin-Fugier, E. (2010). *Bullfighting: a troubled history*. New Jersey: Reaktion Books.
Harper, A. (2011). Connections: Speciesism, racism, and whiteness as the norm. *Sister species: Women, animals, and social justice*, 72–78.
Horta, O. (2017). *Un paso adelante en defensa de los animales*. Madrid: Plaza y Valdés.
Hribal, J. (2003). Animals are part of the working class: a challenge to labor history. *Labor History*, *44*(4), 435–453.
Kalof, L. (2014). Animal blood sport: A ritual display of masculinity and sexual virility. *Sociology of Sport Journal*, *31*(4), 438–454.
Kappeler, S. (1995). Speciesism, racism, nationalism . . . or the power of scientific subjectivity. *Animals and women: Feminist theoretical explorations*, 320–352.

Landborn, A. (2015). *Flamenco and Bullfighting: Movement, Passion and Risk in Two Spanish Traditions*. Jefferson: McFarland.
Lazzarato, M. (2000). *Del biopoder a la biopolítica*. Madrid: Multitudes.
Lennon, J. & Foley, M. (2000). *Dark tourism*. New York: Cengage Learning EMEA.
Leong, P., Joseph, S., & Boulay, R. (2010). Applying constant comparative and discourse analyses to virtual worlds research. *Journal for Virtual Worlds Research*, *3*(1). 3–26.
Lewine, E. (2005). *Death and the sun: A matador's season in the Heart of Spain*. New York: Houghton Mifflin.
Light, D. (2017). Progress in dark tourism and thanatourism research: An uneasy relationship with heritage tourism. *Tourism Management*, *61*, 275–301.
López, E. (2020) "El futuro de la tauromaquia en México, ¿extinción o salvación?" *AS México* [online]. Retrieved May 8, 2020, from https://mexico.as.com/mexico/2019/11/02/reportajes/1572671083_909782.html
Low, P. (2012). *The Cambridge Declaration on Consciousness* [online]. Retrieved October 25, 2019, from http://fcmconference.org/img/CambridgeDeclarationOnConsciousness.pdf
Lozanski, K. (2015). Heroes and Villains: Travel, Risk and Masculinity. In M. Casey (Ed.) *Men, Masculinities, Travel and Tourism* (pp–28–42). London: Palgrave Macmillan.
Markwell, K. (2015). *Animals and tourism: Understanding diverse relationships*. New York: Channel View Publications.
Marvin, G. (1994). *Bullfight*. Chicago: University of Illinois Press.
Medina, L. (2018). Toro Bravo, de muy noble y fiera estampa. *Revista Ibérica* [online]. Retrieved November 2, 2019, from https://www.revistaiberica.com/toro-bravo/.
Mestre, R., Del Rey, A., & Stanishevski, K. (2008). The image of Spain as tourist destination built through fictional cinema. *Journal of Travel & Tourism Marketing*, *24*(2–3), 185–194.
México Taurino (2017). Facebook group site. Retrieved October 14, 2017, from https://www.facebook.com/TauromaquiaTMX.
Miles, W. (2002) Auschwitz: Museum Interpretation and Darker Tourism, *Annals of Tourism Research*, *29*(4), 1175–1178.
Mitchell, T. (1991). *Blood sport: a social history of Spanish bullfighting*. Philadelphia: University of Pennsylvania Press.
Mosterín, J. (2010). *A favor de los toros*. Pamplona: Laetoldi.
Muñoz A., Agüera, E., & Castejón, F. (2007). Muscle glycogen depletion pattern and metabolic response in bulls after bullfighting. *Analecta Veterinaria*, *27*(2), 5–10.
Murray, B. & Muong, V. (2015). "Dirty tourism in Cambodia". *Phnom Penh Post* [online]. Retrieved June 5, 2017, from www.phnompenhpost.com/post-weekend/dirty-tourism-cambodia.
Ortiz, G. (2014). Ética para matador: Savater, los toros y la ética. *Tópicos (México)*, (46), 205–236.
Özlem, A., & Çiftci, G. (2013) Animal Ethics in tourism. *International Journal of Business and Management Studies*, *5*(1) 43–53.
Pezzullo, P. (2009) This is the only tour that sells: tourism, disaster and national identity in New Orleans, *Journal of Tourism and Cultural Change*, *7*(2), 99–114.
Puleo, A. (2014). *Ecofeminismo para otro mundo posible*. Madrid: Ediciones Cátedra.
Purroy, A., & González, J. (1985). Étude des enzimes plamatiques des taureaux de combat tués en corridas. *Reproduction, nutrition, développement*, *25*(4A), 599–603.
Quintero, G. & López, A. (2018). Tauromaquia y turismo oscuro en México: las corridas de toros como prácticas no éticas. *Teoría y Praxis*, *24*, 197–228.
Rangel, N. (2004). *Historia del toreo en Mexico*. Mexico City: Renacimiento.
Regan, T. (1983). *The Case for Animal Rights*. Sacramento: University of California Press.
Saborit, J. (2010). *Concepto Taurino* [online]. Retrieved November 2, 2019, from http://conceptotaurino.blogspot.com/2010/10/.

Salazar, P. (2013). "En México, la tauromaquia es Patrimonio Cultural". *Chilango* [online]. Retrieved October 1, 2017, from http://www.chilango.com/cultura/toreo-sensibilidad-y-arte.

Salleh, A. (1997). *Ecofeminism as Politics Nature, Marx, and the Postmodern*. London: Palgrave.

Savater, F. (2010). *Tauroética*. Madrid: Turpial.

Servitoro (2019) *What is Bullfight?* Retrieved October 28, 2019, from https://www.servitoro.com/en/bullfighting-terms.

Sharpley, R. (2009) Shedding Light on Dark Tourism: An Introduction. In R. Sharpley, R. & P. Stone (Eds.). *The Darker Side of Travel. The Theory and Practice of Dark Tourism* (pp. 3–22) Bristol: Channel View Publications.

Singer, P. (1975). *Animal liberation*. Melbourne: Random House.

Singer, P. (2009). Speciesism and moral status. *Metaphilosophy*, *40*(3–4), 567–581

Stafford, E. (1998). Masculine values, feminine forms: on the gender of personified abstractions. *Thinking Men: Masculinity and Its Self-Representation in the Classical Tradition*, 43–56.

Steinbock, B. (1978). Speciesism and the Idea of Equality. *Philosophy*, *53*(204), 247–256.

Stone, P. (2006). A dark tourism spectrum: Towards a typology of death and macabre related tourist sites, attractions and exhibitions. *Tourism: An Interdisciplinary International Journal*, *54*(2), 145–160.

Stone, P. & Sharpley, R. (2008). Consuming dark tourism: A thanatological perspective. *Annals of tourism Research*, *35*(2), 574–595.

Taurinos Unidos (2017). Facebook group site. Retrieved October 14, 2017, from https://www.facebook.com/groups/orosedasangreysol.

Tauroturismo (2017). TauroTurismo, turismo taurino. Pasión y Arte del toreo. Turismo cultural en íntima combinación con el arte taurino [online]. Retrieved July 8th, 2019, from www.tauroturismo.es/es/quienes-somos.

Thompson, C. & D. Holt. (2004). How do men grab the phallus? Gender tourism in everyday consumption. *Journal of Consumer Culture*, *4*(3), 313–338.

Thompson, K. (2013). Cojones and Rejones: Multiple ways of experiencing, expressing and interpreting gender in the Spanish Mounted Bullfight (Rejoneo). *Gender and Equestrian Sport*, 127–147.

Toros y Toreros (2017). Facebook group site. Retrieved October 14, 2017, from https://www.facebook.com/groups/2399542128.

TripAdvisor (2019a). Plaza de Toros Cancún [online]. Retrieved August 30, 2019, from https://www.tripadvisor.com.mx/Attraction_Review-g150807-d4726995-Reviews-Plaza_de_Toros_Cancun-Cancun_Yucatan_Peninsula.html

TripAdvisor (2019b). Monumental Plaza de Toros [online]. Retrieved August 30, 2019, from https://www.tripadvisor.com.mx/Attraction_Review-g150800-d7697720-Reviews-Monumental_Plaza_de_Toros_Mexico-Mexico_City_Central_Mexico_and_Gulf_Coast.html

TripAdvisor (2019c). Plaza México [online]. Retrieved August 30, 2019, from https://www.tripadvisor.com.mx/Attraction_Review-g150800-d267701-Reviews-Plaza_Mexico-Mexico_City_Central_Mexico_and_Gulf_Coast.html

Urbanik, J. (2012). *Placing animals: An introduction to the geography of human-animal relations*. New York: Rowman & Littlefield.

Urbanik, J. & Johnston, C. (2017). *Humans and Animals: A Geography of Coexistence*. Santa Bárbara: ABC-CLIO.

Van Broeck, A., & López, A. (2018). Turismo oscuro. De la conmemoración a la comodificación de la muerte, los desastres y lo macabro. *Teoría y Práxis*, *24*, 23–68.

Vanda, B., Edwards, C. & Cossío, A. (2017). Dictamen forense sobre el dolor y sufrimiento de los toros durante la corrida, como evidencia de maltrato deliberado. Universidad Nacional Autónoma de México [online]. Retrieved October 30, 2019, from http://www.paot.org.mx/micrositios/animales/pdf/toros/Dictamen_Forense.pdf.

Wilson, J. (1998). *Stephen's Guide to the Logical Fallacies.* Manitoba: Stephen Downes.
Wolfe, C. (2009). Human, All Too Human. *Animal Studies and the Humanities. pmla, 124*(2), 564–575.
Yates, R. (2009). Rituals of dominionism in human-nonhuman relations: Bullfighting to hunting, circuses to petting. Journal of Critical Animal Studies, 7(1), 132–171.

Paul A.G. Tully, Neil Carr

6 Farm animals' participation in tourism experiences: A time for proper respect

Introduction

This chapter analyses farm animals' participation in tourism, looking to change perception of them from simple objects of rural life to active participants in tourism experiences. The next section establishes how farm animals are currently seen as rural objects in tourism via commodification from their agricultural utility. It highlights that both operators and researchers have failed to recognize the sentient nature of them, and, thus, identifies a need for a new tourism research agenda. The chapter then moves to establish why farm animals should be considered active participants in tourism due to them having physical, mental, emotional, and social interests. The discussion uses analysis of farm tourism promotion in New Zealand to illustrate how these intelligent animals participate in tourism experiences. This then allows the chapter to set out how tourism studies can change to provide a better understanding of animals' lived experiences. Therefore, it sets a foundation for a new research agenda by challenging the anthropocentric dominance of tourism and establishing farm animals as active participants. This view is crucial for making sure that animals' physical, mental, emotional, and social lives receive proper respect in the tourism industry.

Contextualizing farm animals and tourism

Farmed animals' commodification in tourism

Each year billions of animals are reared globally on farms to serve human needs. For instance, millions are slaughtered every month so that people can eat meat in countries around the world (see Eurostat, 2018; USDA, 2018). Animals that are farmed and valued by society for their function as consumptive products and those who provide farm resources, such as draught power, live their entire lives affected by human exploitation (Stuart, Schewe, & Gunderson, 2012). In the 21st century, the majority of farm animals are benefiting humanity through factory farms, as technological advancements and capitalism increasingly mechanize processes of their exploitation for human consumption (Peggs, 2012). This growing industrialization of agriculture has resulted in a widening gap between humans and the reality of life for farm animals, as the majority of people continually ignore the origins of their food (Wilkie, 2010).

https://doi.org/10.1515/9783110664058-006

People are conditioned to, instead, think about farm animals through a mythical rural idyll. Based on romantic representations and traditional beliefs, this idyll is presented as what constitutes rural life (Boogaard, Bock, Oosting, & Krogh, 2010). It is made through positive representations of nature, quietness, lush greenery, non-urban, a natural way of life, and rolling hilly landscapes. Within this idyll, wild animals happily live in woods and forests, and farmers care for the domesticated species who roam around farmyards or graze freely in grassy fields (Bell, 2006). This idyll presents these animals as "farm" rather than "farmed" animals. This is an important distinction, with the latter focused on the utilizing of the animal for the benefit of humans, while the former situates the animal as one "living" on the farm. Hence, this rural idyll offers people a sanitized image of farm animal life, and thereby countryside life in general (Yarwood, 2005). Bell (2006) explains how these positive ideas about rurality have become globalized, infiltrating worldwide societies via transnational cultural practices such as music, movies, and the internet, thus offering a global audience an appealing image of rural life. Unsurprisingly, therefore, this image proves to be a popular resource for a rural tourism sector that caters to millions of tourists worldwide and generates billions in economic returns (Lane & Kastenholz, 2015).

The diversification of farm-related businesses into tourism operations relies on selling elements of this rural idyll (Brandth & Haugen, 2012). As such, aspects that connect with positive rurality are commodified for touristic uses, as exemplified by farm animals. Research has established that animals are turned into commercial products by tourism operations (Duffy, 2014). They are exploited, as they become a symbolic resource for attracting people and a physical resource for tourists to experience (Tully & Carr, 2019). In this way, the farm animal of the rural idyll is transformed into a farmed animal. Moreover, tourism operations objectify animals to enhance their ability to create human pleasure and profit (Carr & Broom, 2018). As farms diversify with tourism offerings, animal use is commodified from agricultural utility into a resource of tourism. Across many spatial and cultural contexts, numerous tourism-related activities involving farm animals are taking place. For example, goats can be fed by tourists in Estonia (Kallaste Talu, 2019), chicks are available for cuddles in South Africa (Flag Farm, 2019), cows are milked as part of a live farming demonstration in Canada (Maplewood, 2019), and pigs are raced in Wales (Cantref, 2019). Further, tourists can watch a sheep-shearing demonstration in rural Taiwan (Cingjing, 2019) or visit a city farm in England to meet and greet pigs (Hounslow Farm, 2019). Farm animals are, clearly, an attractive rural resource for the tourism industry.

Research into farm tourism acknowledges that animals play an important role in tourists' experiences. Indeed, Cassel and Pettersson (2015) note that on a Swedish farm, a small collection of animals are kept as "tourism animals" – animals used specifically for the tourists' experience. Further examples of the recognition of the role of farm animals in tourism experiences include, Garrod (2011) highlighting that they are raced for tourist entertainment, and Di Domenico and Miller (2012) noting that activities like cow milking are used to generate an authentic farm feel. Ingram (2002) has also illustrated the educational role of animals interacting with young visitors. However, these studies, as well as others (see Brandth & Haugen, 2011; Talbot, 2013), do not consider these animals as sentient beings and position them as simply an object of rural life. Even the recent wave of critical animal studies in tourism has yet to examine farm animals in tourism to any great extent, with a couple of notable exceptions (see Bertella, 2020; Sayre & Henderson, 2018). Rather, research has focused heavily on the wild, endangered, exotic, and iconic species context and, as a result, has missed the many millions of apparently mundane and unexciting domesticated animals (Carr, 2014; McConnell, 2005) that tourism affects (Young & Carr, 2018). Consequently, recognition of their subjective experiences and, thus, respect for their interests is absent from both research and the global industry.

Farm animals as active participants

Farm animals are not simple objects, but intelligent beings. Animal behavioral research shows that they have active engagement with their surroundings, actions, decisions, and interactions (Broom, 2014; Emel, Johnston, & Stoddard, 2015). Chickens, for example, display a need to investigate their environments and show signs of boredom when it is kept monotonous (Newberry, 1999). Cattle herds operate through an organized social hierarchy in which certain cows exhibit more controlling power over others (Sołtysiak & Nogalski, 2010). Additionally, sheep demonstrate the ability to learn and adapt their behavior in interactions with fellow sheep, inanimate objects, and humans (Kendrick, 2008). These species, and others, demonstrate intelligence to understand the actions of themselves and others, the consequences of an act, an appreciation of their environment and its relevance to their lives, and the aptitude to experience mental states.

These cognitive abilities help to show that farm animals are subjective others as opposed to simple objects. It is clear from an abundance of scientific research that different species have complex cognitive abilities. They have some level of agency or, in other words, the ability for self-willed action (Irvine, 2004). For instance, Bracke (2011) illustrates that pigs consciously roll in mud and water as a way to enhance their life experience via sunburn prevention, cooling, and parasite removal, amongst other functions. It is also now widely recognized that animals, farmed or otherwise, have emotional capacities (Boissy & Erhard, 2014). As an example of this, increased stress has been recorded in goats when they are isolated from their social group (Kannan, Terrill, Kouakou, Gelaye, & Amoah, 2002). Farm animals are also conscious individuals who have a "capability to perceive and respond to sensory stimuli" (Broom, 2014, p. 76). This has been shown by Marino and Allen (2017) in cows, who display a strong reliance on the five senses of sight, hearing, touch, taste, and smell in daily life. This discussion provides insight into the complex cognition of farm animals. Consequently, it is important to recognize that they participate in tourism experiences not simply as objects of rural life but as living beings with multifaceted physical, mental, emotional, and social lives.

This intelligence demonstrates the sentient natures of farm animals (Broom, 2014). Recognizing sentience means accepting that animals have interests – physical, mental, emotional, and social needs and wants – in how they live their lives. How humans infringe on these interests should be of concern to us as we (humans) move through a time of increasing animal respect in the public's social conscience (Carr & Broom, 2018). Sentience creates implications for human action towards animals. Specifically, its recognition leads to a moral status that generates obligations for responsible human action (Broom, 2014). In other words, accepting that animals have subjective lives should result in actions towards how they are treated. However, missing from tourism research and the wider industry is recognition of farm animals' sentience and an understanding of the obligations we owe to their lives. This is an important gap, given both the scale of tourism activity involving farm animals and increasing concern around animal welfare in the tourism industry.

Moving beyond animal welfare in tourism

Critical animal studies in tourism research, predominantly, considers the human-driven concepts of animal welfare and animal rights. However, animal welfare, which is socially viewed as the reduction of potential animal suffering during human use, receives substantially more attention than the rights concept (Fennell, 2015). Eliminating people's abuse and neglect of animals, providing better environments for animals held captive by humans (i.e. bigger cages), and ensuring human handling of animals is done to a correct skill level would be outcome examples to satisfy welfarists (Grandin, 2015). Proponents of a hard animal rights approach, on the other hand, pursue an abolitionist agenda for the use of animals by humans (Fennell, 2015). Their quest for abolishment is often seen as controversial and fanciful and, as such, is perceived as a fringe movement of society associated with disruptive protests, demonstrations, and other confrontational actions (Peggs, 2012; Sunstein, 2004). This is despite many variations on the animal rights concept now contributing to key legislative and educational debates in society (DeMello, 2012). Quite possibly a deviant labeling of hard animal rights approaches in society may have played a role in the dominance of animal welfare as the criteria for assessing humans' use of animals.

The principal animal welfare concentration in tourism indicates, seemingly, an acceptance with using animals, providing concern for their well-being exists (Sheppard & Fennell, 2019). However, it is questionable if simply having a concern for well-being is a satisfactory criterion for justifying animal use, as the human notion of well-being is often too simplistically described in society as "the state of feeling healthy and happy" (Cambridge Dictionary, 2019). Instead, well-being should relate to a multilayered concept that involves mental and physical health aspects intertwined with ideas of happiness, enjoyment, engagement, fulfillment, and satisfaction gained from one's environment and actions (Haworth & Hart, 2007). Given the recognition of the multifaceted physical, mental, emotional, and social lives animals live, discussions outside of tourism have increasingly brought attention to the belief that "good welfare really isn't good enough", as it still means animals suffer mistreatment and disrespect at human hands (Bekoff, 2008, p. 3). Zoo animals, for example, often receive the highest quality of care but are still placed into unusual environments, separated from family members, made to interact unnaturally with people, and commodified to generate human rewards (Henderson, 2011; Mellor, 2016). Hence, whilst ensuring high well-being standards is a critical component for an animal's life, it does not stop them from being constrained, manipulated, deprived, and enslaved by human use. Therefore, research must show that respect for animal life is not only about well-being but requires consideration for the complete life lived by the species involved. In adopting this approach, tourism would be best served by moving away from the human beliefs that underpin animal welfare and animal rights to a position that actively seeks to uncover what an animal truly needs and wants and how this information should influence the use of animals by humans (Carr & Young, 2018). Only through hearing the animal "voice" can we be in a position to bring proper respect to the farm animals that are a part of the tourism experience.

Methodology: Exploring farm animals' participation in tourism

To help build a new research agenda for farm animals as part of the tourism experience, the remainder of this chapter uses data from an exploratory study of farm attraction websites. Data from 29 farm attraction websites in the New Zealand tourism sector were collected in April 2019 and analyzed to uncover animals' tourism participation. In the 21st century, the internet has become a substantial data source for researchers looking to explore aspects of leisure (Ward, Clark, Zabriskie, & Morris, 2014). At the same time, the online environment has grown to exert considerable influence over perceptions of tourism experiences (Choi, Hickerson, & Kerstetter, 2018). Hence, web-based studies have increasingly become a widely used and important data collection method in tourism research (Mkono & Tribe, 2017). Whilst as a data source it has limitations in what can be uncovered (see Ward et al., 2014) – such as the inability to position each attraction on a farm-tourism continuum from those farms that remain predominantly agricultural to those that are now fully dependent on tourists (Busby & Rendle, 2000) – online exploration does enable, as will be seen, the ability to highlight aspects of farm animals' participation in tourism for future investigation.

To begin the research process, the first author used the influential industry website TripAdvisor to search "farm attractions" in New Zealand. Each result was then studied and 29 websites were selected for in-depth analysis. Four specific aspects were considered in this selection process – did TripAdvisor mark them with a "featuring animals" icon, was a website link for an attraction provided, had visitors left reviews of the attraction, and was clear evidence available of farm animal use. Once these criteria were met, the websites linked to each attraction were bookmarked ready to undergo a qualitative content analysis. This interpretive style of analysis relies on a researcher studying the array of content forms – texts, pictures, videos, animations, and other graphics – that make up online environments (Kim & Kuljis, 2010). Accordingly, the first author critically analyzed each page of the websites by studying textual and visual messages about farm animals. For example, he read about the available animals at attractions, viewed picture galleries of tourist-animal selfies, watched videos of human-animal stage shows, and noted headlines and blurbs about animal use. For each website, the information relating to farm animals in tourism operations was manually extracted into individual Microsoft Word documents.

This enabled the first author to conduct a detailed analysis of the data. These 29 Word documents were printed and re-read in order to gain familiarity with the data and, thus, uncover emergent themes. As themes emerged the documents were read multiple times, along with color-coding and annotations, until all possible considerations had been exhausted. As such, the analysis process followed a thematic approach, as the first author looked to unpack the implicit and explicit "ideas within the data" (Guest, MacQueen, & Namey, 2012, p. 10). This enabled the data to be thoroughly analyzed, as well as giving a focus to uncovering the deeper meanings within the rich dataset (Walters, 2016). In other words, the first author looked to infer the messages communicated to tourists about animals in the text and visual composition of websites. It is important to recognize in this context that when undertaking this analysis process the intent was to see "beyond the human" and to begin to understand the material presented from the perspective of the farm animals. As Silverman (1997, p. 171) points out "putting oneself 'in the animal's place' [to try to understand them] is a formidable task." Yet, doing so may actually be viewed as being no different from understanding the human other, difficult but not impossible (Bekoff, 2007; Carr, 2014).

The first author organized the coded data into five categories: imagery, animals' welfare, tourist-animal interaction, tourism-related activities, and anthropomorphized animals. The data was transferred to new Word documents representing each category with each piece of evidence given a reference to its website origin. This allowed further scrutiny of the material and theoretical connections to be made. This analysis contributes in the remainder of this chapter to the creation of a new research agenda for farm animals' participation in tourism.

Analysis

Farm animals' participation in tourism

Farm animals should be considered as active participants in tourism experiences. One driver of this is the recognition of these animals as sentient beings, while additionally, it can be driven by the recognition that tourists are increasingly searching for interaction and participation with all components of their experiences (Campos, Mendes, Oom do Valle, & Scott, 2017). Based on the latter, website marketing communicates the idea that interaction with animals can be a central part of farm tourism experiences. For instance, impactful images of a young sheep with eyes closed rubbing against a human hand or happy-looking children captured having the chance to feed various species build beliefs about the role of animals on the farm. Likewise, text – such as "anything from lambs to a Clydesdale, from guinea pigs to llamas, and much more can be patted, fed and watched" (Glenorchy, 2019), "best sheep selfie in New Zealand" (Real Country, 2019), and "get a real 'hands-on' experience with the animals" (Hawkes Bay, 2019) – says to tourists that animal interaction is encouraged. Indeed, 27 of the 29 farms aim to capitalize on this demand, including promoting the chance to undertake non-working farm activities like cuddling, hand feeding, riding, playing with, petting, or posing with animals. Additionally, at a number of the farms, the opportunity for involvement with traditional farm undertakings like sheep shearing, cow milking, and sheep herding is promoted. This marketing strategy communicates two important perceptions to the tourists, one about the notion of animals as an object of the rural idyll and one for seeing them as an acceptable resource of tourism. Objectifying farm animals in this way, though, uses them as an instrument for creating human pleasure and profit, disrespecting their value as sentient beings. However, as sentient beings farm animals are active participants in these tourism experiences. These interactions will, in currently unknown ways, affect the physical, mental, emotional, and social lives of these animals.

The growing diversification of farming life into tourism attractions is resulting in people cuddling lambs, sitting on cows, and tickling pigs. All these depictions on marketing websites look to create two important perceptions for the tourists, one about the notion of animals as an integral object of the rural idyll and one for seeing them as an acceptable resource of tourism. Farms develop these tourism experiences driven by the need to create human entertainment and profit maximization. Even the traditional "work" activities of farm animals are commoditized for tourism gains, as activities like sheep shearing become a human-animal stage show performed for tourists (Figure 6.1). In this performance, though, where is the respect when a sentient sheep, already stressed by the shearing experience (Hargreaves & Hutson, 1990) and with visible resistance, is manhandled onto a stage in front of a noisy crowd who laughs, claps, and cheers? Do tourists' see a knowledgeable human performer (aka the sheep shearer) and accept the show as fine? This would fit with the concept of "diffusion of responsibility," the idea that the presence of others reduces one's sense of responsibility in social situations (Beyer, Sidarus, Bonicalzi, & Haggard, 2017). Ultimately, the commoditization of the sheep into a source of entertainment has the potential to induce a disregard for an animal's sentience in the minds of tourist and/or operator. They are positioned as an object of the farm via the rural idyll and cared for by a knowledgeable farmer. Thus, they are free to be enslaved in the human leisure experience. Just as people are conditioned not to notice the animals' presence in their use in the human food chain, the animals can also seemingly be made invisible in human entertainment. This fits with work on zoos that has questioned the ability of entertainment-oriented zoos to raise awareness of animal welfare among visitors (Carr & Broom, 2018).

Figure 6.1: A sheep shearing show.
Source: Sheepworld (2019). Used with permission.

A need exists in tourism research to explore how animal interests fit these anthropocentrically dominated experiences. Does thought for the lives of farm animals play any part in their creation, design or undertaking? Do these interactions satisfy the animals' needs and wants? Is it possible for these interactions to benefit both human and animal participants? Only by treating their sentient natures respectfully, can future research answer these types of questions and truly uncover the impact of tourism participation on farm animals' physical, mental, emotional, and social lives. Addressing these issues must be central to the development of farm animal tourism experiences based on the idea of co-creation, the notion that multiple active subjects form tourism experiences (Bertella, 2014), between animals, tourists, and operators.

Farm animals' manipulated reality

A critical component in farm animals' invisibility is how marketing discourse manipulates the tourists' view away from the reality of animal life. For instance, promotion concentrates on making the animals relatable to people by using the human-constructed category of pet. Whilst a few farms directly refer to their animals as pets, most take advantage of the embedded societal connotations that people have about this category. The people-pet relationship is one thought about via ideas of love and affection, as the friendly and gentle pet is welcomed into the caring humans' home (Peggs, 2012). In reality, the people-pet relationship can be said to be one of invisible human dominance as people treat animal life in ways that generate their personal pleasure and satisfaction (Tuan, 1984). A part of this control involves people providing their pet with pseudo-human status through anthropomorphic symbols (Fudge, 2008), which are things that are associated with humankind such as names and traits. This anthropomorphic construction of animals is also a feature of creatures that inhabit the rural idyll. Hence, farm marketing aims to capitalize on the pet category and representations of the rural idyll by naming their animals (e.g., Frankie the cow, Chester and Ernie the baby goats, and Penelope the pig). Additionally, attributing human traits to animals, such as "the inquisitive goat" (Farm Encounters, 2019) and "play with all our cuddly little friends" (Agrodome, 2019), helps to reinforce the pet connection. None of these animals are pets, though. Instead, this connection is designed to mask the reality of these animals being forced to "work" by humans as tourism (and farm) laborers. In doing so, it assists in obscuring the fact that farm animals' lives are dominated by human exploitation.

An element of attractiveness is added to farm animals in website marketing. The reality of farms as a place of muck and dirt (Hamilton, 2007) is often removed in favor of messages about photo-ready, clean, and groomed animals. This is done to generate positive perceptions from tourists' regarding the use and treatment of animals and, in turn, make them more appealing. Also more appealing are animals displayed in wide-open spaces with picturesque green landscapes as a backdrop. Thus, suggesting that farm animals are naturally at home in and free to roam the rural environment. These presentation facets – of farm animals as attractive, natural, and free – obscure how these animals are worked, controlled, and kept by humans. As well as being clean, the animals are healthy. Death and dying, a reality of life, especially on the farm, are not present in the imagery presented to tourists. Old, disabled animals are absent, as are depictions of the transformation of farm animals into meat products, the raison d'être for most of them. These depictions, undoubtedly, aim to benefit from beliefs about the rural idyll and at the same time reinforce them. As Yarwood (2005) suggests, for farm animals to be seen as acceptable for tourism operators use a sanitized image of them is specially constructed for the tourist gaze. Thus, as is a common theme across tourism sectors, farm animal imagery becomes the idealized rather than the real (MacCannell, 2011). This manipulates the reality of farm animal life, presenting them in a photogenic manner that supports their exploitation for anthropocentric tourism gains with a seeming disregard for their needs and wants.

Photographic elements also add an anthropomorphized outcome to the animals' presentation. People are increasingly searching for opportunities to take a selfie and gain social media recognition in their leisure activities (Pearce & Moscardo, 2015). Marketing, thus, aims to appeal to this desire through wording that supports the photographic potential of farm experiences, such as, "full of attitude, seriously cute and very photogenic – come and meet our friendly alpacas" (Shamarra, 2019) and "our lambs are very friendly, cuddly, and they love being part of the show and are very photogenic" (The Point, 2019) (Figure 6.2). This text displays the animal as a willing

Figure 6.2: Websites promote the photographic potential of farm tourism experiences. Source: Shamarra (2019). Used with permission.

participant in the human act of photography. One farm is even explicit in proclaiming that their animals' love having a picture taken with tourists. Building this pseudo-human status around farm animals, both through the connection with the pet category and this social trend for photography, helps to mask their sentient reality and, consequently, disrespect the interests they have in the lives they live. In shedding light on farm animals' physical, mental, emotional, and social lives in tourism experiences, research can be in a position to help the global sector understand the true reality of their tourism participation.

Concern for farm animals' welfare

Farm animals' manipulated presentation continues with ideas about well-being. Only one website displayed any concern for the actual animal welfare principle of reducing animals' potential suffering, with text warning tourists of the need to "use quiet voices around animals, loud and unexpected noises can give them frights" and "be gentle with the animals in the nursery shed, as they are small and can get injured easily from rough handling" (Kiwi Valley, 2019). These messages place care for the animals lived experience as a focal aspect of their welfare concern. However, welfare discourse on the majority of farm websites focuses not on ways to reduce experienced animal suffering but concentrates, instead, on tourist satisfaction. Animal welfare appears to be used as a marketing resource with, for instance, text proclaiming, "they are happy and well-cared for, so are fun to meet" (Bullswool, 2019) and "gorgeous, friendly, healthy and happy animals" (Funky Farm, 2019). These health and happiness narratives focus not on respect for the animals' lived experience but on using the positive and simplistic societal beliefs around good well-being to support anthropocentric tourism gains.

Establishing proper respect for farm animals in tourism involves taking their lives seriously. This requires removing the anthropocentrism of tourism, which sees outcomes of human pleasure and profit dominate animals' sentient natures and views good animal well-being as an acceptable welfare standard. Tourism as a whole must be clear that seeing simply good health and happiness does not equate to good welfare (Buller, Blokhuis, Jensen, & Keeling, 2018). As Tully and Carr (2019) discuss, animals can appear healthy and content in their actions but still suffer a wide range of negative consequences courtesy of the physical, mental, emotional, and social infringements that humans inflict on them. Sentient animal species have, like humans, complex brain systems that means poor welfare can occur in many ways (Broom, 2014). Hence, welfare is not achieved merely by assessing human criteria of well-being but is about the animals "experienced, lived and embodied reality" (Bock & Buller, 2013, p. 391). Therefore, taking their lives seriously requires an approach that is undertaken with empathy for the complete life lived by an animal. This begins by understanding that farm animals are active participants, albeit not of their own volition, of tourism and treating their lived experience respectfully.

Discussion: A need to fulfill human obligations

To realize this respect requires responsible human action. Broom (2014, p. 13) contends that animals are owed certain obligations by humans, which include a "duty to act or to refrain from acting" in ways that adversely affect the animal's life. He bases his belief on developing scientific knowledge that shows that both humans and animals share components of sentience, and, as a result, just as we are obliged to act in responsible ways due to the effects on our fellow sentient humans (Kuper, 2005), our actions towards animals should be about fulfilling human responsibilities. In the same way, for example, that it is wrong to perform an act that causes deliberate injury to a fellow human, it is wrong when an animal is involved. Hence, obligations exist towards farm animals concerning respect for their well-being during all aspects of their lives, including tourism use. So there is a responsibility to refrain from causing physical abuse, and a responsibility to provide access to proper nutrition, for example. Achieving these responsibilities regarding animal well-being does not equal complete respect for an animal's lived experience, as many more obligations exist. At the heart of these obligations is the recognition of a power imbalance. The power rests with humans who may, and have, with impunity abuse this power to ensure their personal well-being and titillation at the expense of animals. Following Broom's and others (see May, 2014) view about obligations, there is a need for humans to recognize this power imbalance, take responsibility for it, and not abuse it.

This chapter uncovered potential failings in certain obligations owed to farm animals. Such as, when a cow is anthropomorphized into "Frankie" and becomes a "tourism animal," does tourism activity interfere with the social life found in her herd? Is anthropomorphizing an animal into a pseudo-human tourist attraction an infringement of their dignity? If human responsibility includes enabling fellow humans to fulfill social needs without humiliation (Pogge, 2005), why should this not apply to the animals of tourism? Additionally, commoditizing a sheep via a stage performance of an activity that they find stressful is wrong. Humans are obliged not to cause mental suffering in another human (Andorno & Baffone, 2014). Why is it right, therefore, to accept a sheep's mental suffering in the quest for tourism pleasure and profit? Furthermore, it is wrong for humans to enslave another into a situation that gives them a lack of control and no ability to achieve personal interests (Stevenson, 2015). So why is it acceptable to enslave animals for tourism gains? Farm animals have no choice about participating in tourism, thus is it not time to abolish such animal enslavement? This is especially the case as we fail to respect their interests in so many ways. Only through addressing these human responsibilities and linking further obligations with tourism's animal use will proper respect for farm animals be possible.

Fulfilling any obligation is made difficult by the depiction of farm animals in tourism. As discussed, the reality of farm animal life is manipulated, conditioning tourists into accepting their tourism use without recognizing its exploitative nature. This objectification presents the tourist with a passive object with no interest in the experience, as opposed to a sentient animal with individual consciousness; an instrumental valuing that is found in the tourism promotion of many, diverse species (e.g., donkeys [Tully & Carr, 2019]; elephants [Duffy, 2014]; whales [Cloke & Perkins, 2005]; polar bears [Yudina & Grimwood, 2016]). Promotional messages hold enormous sway in tourism for shaping the perceptions tourists have about animals (Bertella, 2013). Therefore, to generate greater respect for all animal species used in tourism, not just those farmed, the promotional manipulation of their reality must end. Messages that position animals as an instrument of tourism must be replaced. The depiction of animals in this anthropocentrically dominated industry that results in the failure of human obligations and in disrespect for their sentience must change.

To instigate this change, a new way of seeing animals that challenges the anthropocentric power of tourism needs to be embraced. As Bock and Buller (2013) illustrate, the scientific study of animals' welfare has changed substantially as ethological knowledge has grown. New approaches are supporting the notion that animal life is lived not through defined assessment criteria but, rather, a multi-dimensional system of subjective experiences (Buller et al., 2018). Discussions concerning the welfare of animals in tourism need to follow these developments, with a focus on how the industry affects the complete animal life. Tourism must embrace an empathetic approach to animal lives to appreciate the wide range of interests species have in the lives they live. An animal-centric style lens that can advance a research agenda concentrating on the obligations owed to animals during tourism activity has significant potential (Carr & Broom, 2018). In particular, it can radically alter the view of animal treatment in the industry by considering their sentient natures and using this knowledge to challenge the anthropocentrism that controls animal use. Thus, critiquing the failures in obligations owed to farm animals can make visible the changes that are needed to show respect to all animals involved in tourism experiences. In this way, if we are brave enough, tourism really can be an educational tool that is aligned with increasing social concern for animal welfare. This requires us in the case of farm animals to strip away the veneer of the rural idyll and the "disneyfication" of them and encourage tourists to think. In this way, tourism can seek to be a transformational journey rather than just a site of banal entertainment.

Conclusion: A new research agenda

This chapter has established the need for a new research agenda on farm animals in tourism experiences. It has explained that farm animals' lives contain complex physical, mental, emotional, and social interests and that they have the potential to be involved as active participants in tourism experiences. However, the tourism sector has so far treated them as simple rural objects (farmed animals) and, thus, ignored their sentient reality. As discussed, a wide variety of farm animals are increasingly participating in tourism experiences that are either non-agricultural (e.g., cuddling, posing with, racing) or traditionally agricultural (e.g., sheep shearing, cow milking). Across spatial and cultural contexts, farm animals are used for tourist entertainment and the generation of operator profits. Therefore, studies are needed from the rural farms of Taiwan to the city farms of England, and everywhere in between. This research should focus on farms at different positions on the farm-tourism continuum, from those that remain largely agriculturally based to the fully converted visitor attractions, thus allowing for an understanding of how the extent of tourism participation affects the lives of these animals. A new research agenda that employs an animal-centric lens to shed light on the reality of animals as co-creators in the farm tourism experience is needed. What is required, moreover, is an approach that is undertaken with empathy for the complete life lived by farm animals. Only by making visible the reality experienced by all species of farm animal affected by tourism, both directly and indirectly, can we begin to offer proper respect to them. This is the route to the realization of true co-creation and the transformation of the farmed animal as object into the farm animal as sentient being.

Within fluid tourism spaces, where definitions of farms and farm animals are being re-imagined, it is important to uncover what happens to the animals' interests – their physical, mental, emotional, and social needs and wants – during tourism experiences. We need to question how tourism infringes on animal interests and understand the relevant consequences for their lives. We need to ask, if human-animal interactions in the tourist space have any potential to benefit the latter, at the individual and/or species level. It is also important to consider the farm animal outside of tourism. What does keeping "tourism animals" mean for the treatment of all animals on a farm? Does being a "tourism animal" result in special treatment for particular animals over others? Does tourism have any impact on or responsibility for those kept behind the scenes for agricultural utility? Additionally, studies need to investigate the human participants, both the tourists and operators, by considering the current and possible future roles of animal sentience in their experience. Furthermore, using knowledge about their sentient natures, scholars should study how the tourists are conditioned into seeing a rural object as opposed to a sentient animal. What is needed, more specifically, is research that influences changes that give precedence to the animals' sentience over tourism's anthropocentric gains of pleasure and profit. In the quest for these gains, tourism deprives, constrains, manipulates, and enslaves farm animals, which impacts on their physical, mental, emotional, and social lives. On-farm studies are required to investigate the accomplishment or failure of the many obligations humanity should have for all sentient animals, even those seemingly as mundane as farm animals. Uncovering these obligations can lead to a realization in tourism that animals' welfare is not only about seeing good animal well-being but requires consideration of how human activity infringes on all their interests.

References

Agrodome (2019). Agrodome Farm Experience. Retrieved from: https://www.agrodome.co.nz/

Andorno, R. & Baffone, C. (2014). Human rights and the moral obligation to alleviate suffering. In R. Green and N. Palpant (eds.) Suffering *and bioethics* (pp. 182–200). New York: Oxford University Press.

Bekoff, M. (2007). *Animals Matter*. Boston, Massachusetts: Shambhala.

Bekoff, M. (2008). Why 'good welfare' isn't 'good enough': minding animals and increasing our compassionate footprint. *Annual Review of Biomedical Sciences*, 10, 1–14. Retrived from: http://arbs.biblioteca.unesp.br

Bell, D. (2006). Variations on the rural idyll. In P. Cloke, T. Marsden, & P. Mooney (Eds.) *Handbook of rural studies* (pp. 149–160). London: Sage Publications.

Bertella, G. (2013). Ethical content of pictures of animals in tourism promotion. *Tourism Recreation Research*, 38(3), 281–294. doi: 10.1080/02508281.2013.11081754

Bertella, G. (2014). The co-creation of animal-based tourism experiences. *Tourism Recreation Research*, 39(1), 115–125.

Bertella, G. (2020). Animal-based experiences and animal experiences: farm animals' perspective on human leisure in rural settings. *Annals of Leisure Research*. doi: 10.1080/11745398.2020.1740603.

Beyer, F., Sidarus, N., Bonicalzi, S. & Haggard, P. (2017). Beyond self-serving bias: diffusion of responsibility reduces sense of agency and outcome monitoring. *Social Cognitive and Affective Neuroscience*, 12(1), 138–145.

Bock, B. & Buller, H. (2013). Healthy, happy and humane: evidence in farm animal welfare policy. *Sociologia Ruralis*, 53(3), 390–411. doi: 10.1111/soru.12011

Boissy, A. & Erhard, H.W. (2014). How studying interactions between animal emotions, cognition, and personality can contribute to improve farm animal welfare. In: T Grandin & M.J. Deesing (eds.), *Genetics and the behavior of domestic animals* (pp. 81–113). London: Elsevier.

Boogaard, B.K., Bock, B.B., Oosting, S.J. & Krogh, E. (2010). Visiting a farm: an exploratory study of the social construction of animal farming in Norway and the Netherlands based on sensory perception. *International Journal of Sociology of Agriculture & Food*, 17(1), 24–50.

Bracke, M.B.M. (2011). Review of wallowing in pigs: description of the behavior and its motivational basis. *Applied Animal Behaviour Science*, 132(1–2), 1–13. doi: 10.1016/j.applanim.2011.01.002

Brandth, B. & Haugen, M.S. (2011). Farm diversification into tourism: implications for social identity? *Journal of Rural Studies*, 27, 35–44. doi: 10.1016/j.jrurstud.2010.09.002

Brandth, B. & Haugen, M.S. (2012). Farm tourism and dilemmas of commercial activity in the home. *Hospitality & Society*, 2(2), 179–169. doi: 10.1386/hosp.2.2.179_1

Broom, D.M. (2014). *Sentience and animal welfare*. Wallingford: CAB International.

Buller, H., Blokhuis, H., Jensen, P. & Keeling, L. (2018). Towards farm animal welfare and sustainability. *Animals*, 8, 81–94. doi: 10.3390/ani8060081

Bullswool (2019). Bullswool farm heritage park. Retrieved from: https://www.bullswoolfarm.co.nz/farmpark/

Busby, G. & Rendle, S. (2000). The transition from tourism on farms to farm tourism. *Tourism Management*, 21(6), 635–642. doi: 10.1016/S0261-5177(00)00011-X

Cambridge Dictionary (2019). Online Cambridge dictionary edition. Retrieved from: https://dictionary.cambridge.org/dictionary/english/well-being?q=wellbeing

Campos, A.C., Mendes, J., Oom do Valle, P. & Scott, N. (2017). Co-creating animal-based tourist experiences: attention, involvement and memorability. *Tourism Management*, 63, 100–114. doi: 10.1016/j.tourman.2017.06.001

Cantref (2019). Pig racing. Retrieved from: https://www.cantref.com/pig-racing-at-cantref-adventure-farm

Carr, N. (2014). *Dogs in the leisure experience.* Wallingford, UK: CABI.

Carr, N. & Broom, D.M. (2018). *Tourism and animal welfare.* Wallingford: CABI.

Carr, N. & Young, J. (2018). Wild animals and leisure: an introduction. In N. Carr & J. Young (eds.) *Wild animals and leisure: rights and wellbeing* (pp. 1–11). Abingdon: Routledge.

Cassel, S.H. & Pettersson, K. (2015). Performing gender and rurality in Swedish farm tourism. *Scandinavian Journal of Hospitality and Tourism*, 15(1–2), 138–151.

Choi, Y., Hickerson, B. & Kerstetter, D. (2018). Understanding the sources of online travel information. *Journal of Travel Research*, 57(1), 116–128. doi: 10.1177/0047287516683833

Cingjing (2019). Qingjing sheep shearing show. Retrieved from: https://www.cingjing.gov.tw/en/grassland/index.php?index_id=25

Cloke, P. & Perkins, H.C. (2005). Cetacean performance and tourism in Kaikoura, New Zealand. *Environment and Planning D: Society and Space*, 23, 903–924. doi: 10.1068/d57j

DeMello, M. (2012). *Animals and Society.* New York: Columbia University Press.

Di Domenico, M. & Miller, G. (2012). Farming and tourism enterprise: experiential authenticity in the diversification of independent small-scale family farming. *Tourism Management*, 33(2), 285–294.

Duffy, R. (2014). Interactive elephants: nature, tourism and neoliberalism. *Annals of Tourism Research*, 44, 88–101. doi: 10.1016/j.annals.2013.09.003

Emel, J., Johnston, C.L. & Stoddard, E.L. (2015). Livelier livelihoods: animal and human collaboration on the farm. In K. Gillespie & R-C. Collard (eds.) *Critical animal geographies* (pp. 164–183). London: Routledge.

Eurostat (2018). Agricultural production – livestock and meat. Retrieved from: https://ec.europa.eu/eurostat/statistics-explained/index.php?title=Agricultural_production_-_livestock_and_meat&oldid=427096

Farm Encounters (2019). Close Encounters Farm Experience. Retrieved from: https://www.farmencounters.co.nz

Fennell, D.A. (2015). The status of animal ethics research in tourism: a review of theory. In K. Markwell (ed.) *Animals and tourism: understanding diverse relationships* (pp. 27–43). Bristol: Channel View.

Flag Farm (2019). Flag animal farm gallery. Retrieved from: http://flaganimalfarm.co.za/gallery/

Fudge, E. (2008). *Pets.* London: Routledge.

Funky Farm (2019). Funky Farm. Retrieved from: https://www.funkyfarm.co.nz/pettingzoo

Garrod, B. (2011). Diversification into farm tourism: case studies from Wales. In A. Fyall & B. Garrod (eds.) *Contemporary cases online.* Oxford: Goodfellow Publishers. Retrieved from: https://www.goodfellowpublishers.com/free_files/file12farm.pdf

Glenorchy (2019). Glenorchy animal experience. Retrieved from: http://glenorchyanimalexperience.co.nz/

Grandin, T. (2015). *Improving animal welfare: a practical approach.* 2nd Ed. Wallingford: CABI.

Guest, G., MacQueen, K.M. & Namey, E.E. (2012). *Applied thematic analysis.* Thousand Oaks: Sage Publications Inc.

Hamilton, L. (2007). Cultural transformations in the world of farm animal veterinary surgeons. *Ethnography*, 8(4), 485–501. doi: 10.1177/1466138107083564

Hargreaves, A.L. & Hutson, G.D. (1990). The stress response in sheep during routine handling procedures. *Applied Animal Behaviour Science*, 26(1–2), 83–90. doi: 10.1016/0168-1591(90)90089-V

Hawkes Bay (2019). Hawkes Bay Farmyard Zoo. Retrieved from: http://www.farmyardzoo.co.nz/

Haworth, J. & Hart, G. (2007). Introduction. In J. Haworth & G. Hart (eds.) *Well-being: Individual, community and social perspectives* (pp. 1–24). New York: Palgrave MacMillan.
Henderson, J.C. (2011). Singapore zoo and night safari. In W. Frost (ed.) *Zoos and tourism: conservation, education, and entertainment?* (pp. 100–111). Bristol: Channel View.
Hounslow Farm (2019). Hounslow Urban Farm. Retrieved from: https://hounslowurbanfarm.co.uk/animals/
Ingram, G. (2002). Motivations of farm tourism hosts and guests in the South West Tapestry Region, Western Australia: a phenomenological study. *Indo-Pacific Journal of Phenomenology*, 2(1), 1–12.
Irvine, L. (2004). A model of animal selfhood: Expanding interactionist possibilities. *Symbolic Interaction*, 27(1), 3–21.
Kallaste Talu (2019). Farm animal park. Retrieved from: https://kallastetalu.ee/elamustalu_tegevused/farm-animals-park/?lang=en
Kannan, G., Terrill, T.H., Kouakou, B., Gelaye, S. & Amoah, E.A. (2002). Simulated pre-slaughter holding and isolation effects on stress responses and live weight shrinkage in meat goats. *Journal of Animal Science*, 80(7), 1771–1780. doi: 10.2527/2002.8071771x
Kendrick, K.M. (2008). Sheep senses, social cognition and capacity for consciousness. In C. Dwyer (ed.) *The welfare of sheep* (pp. 135–157). Dordrecht: Springer.
Kim, I. & Kuljis, J. (2010). Applying content analysis to web-based content. *Journal of Computing and Information Technology*, 18(4), 369–375. doi: 10.2498/cit.1001924
Kiwi Valley (2019). Kiwi Valley Farm Park. Retrieved from: http://www.kiwivalley.co.nz./
Kuper, A. (2005). Introduction: the responsibilities approach to human rights. In A. Kuper (ed.) *Global responsibilities: who must deliver on human rights?* (pp. ix–xxii). Abingdon: Routledge.
Lane, B. & Kastenholz, E. (2015). Rural tourism: the evolution of practice and research approaches – towards a new generation concept. *Journal of Sustainable Tourism*, 23(8–9), 1133–1156.
May, T. (2014). Moral individualism, moral relationalism, and obligations to non-human animals. *Journal of Applied Philosophy*, 31(2), 155–168. doi: 10.1111/japp.12055
MacCannell, D. (2011). *The ethics of sightseeing*. Berkeley: University of California Press.
Maplewood (2019). Maplewood Farm. Retrieved from: http://maplewoodfarm.bc.ca/animals/cows/
Marino, L. & Allen, K. (2017). The psychology of cows. *Animal Behavior and Cognition*. 4(4), 474–498. doi: 10.26451/abc.04.04.06.2017
McConnell, P. (2005). *For the love of a dog: understanding emotion in you and your best friend*. New York: Ballantine Books.
Mellor, D. (2016). Updating animal welfare thinking: moving beyond the 'five freedoms' towards a 'life worth living'. *Animals*, 6(3), 21–41. doi: 10.3390/ani6030021
Mkono, M. & Tribe, J. (2017). Beyond reviewing: uncovering the multiple roles of tourism social media users. *Journal of Travel Research*, 56(3), 287–298. doi: 10.1177/0047287516636236
Newberry, R.C. (1999). Exploratory behaviour of young domestic fowl. *Applied Animal Behaviour Science*, 63(4), 311–321. doi: 10.1016/S0168-1591(99)00016-7
Pearce, J. & Moscardo, G. (2015). Social representations of tourist selfies: new challenges for sustainable tourism. In *Conference Proceedings of BEST EN Think Tank XV* (pp. 59–73). From: BEST EN Think Tank XV, 17–21 June 2015, Skukuza, Mpumalanga, South Africa. Retrieved from: https://researchonline.jcu.edu.au/40604/
Peggs, K. (2012). *Animals and sociology*. Basingstoke: Palgrave MacMillan.
Pogge, T.W. (2005). Human rights and human responsibilities. In A. Kuper (ed.), *Global responsibilities: who must deliver on human rights?* (pp. 3–37). Abingdon: Routledge.
The Point (2019). The Point Sheep Shearing. Retrieved from: http://pointbnb.co.nz/sheep-shearing/

Real Country (2019). Real Country Farm. Retrieved from: http://www.realcountry.co.nz/pet-and-play.html

Sayre, R. & Henderson, K. (2018). The cow goes "moo": farm animal and tourist interactions on Long Island's North Fork. In C. Kline (ed.) *Animals, food and tourism* (pp. 82–95). Routledge, Abingdon.

Shamarra (2019). Shamarra Alpaca Farm. Retrieved from: https://www.shamarra-alpacas.co.nz/#tour-section

Sheppard, V.A. & Fennell, D.A. (2019). Progress in tourism public sector policy: toward an ethic for non-human animals. *Tourism Management*, 73, 134–142. doi: 10.1016/j.tourman.2018.11.017

Sheepworld. (2019). Sheepworld - shows & times. Retrieved from: https://www.sheepworldfarm.co.nz/shows-and-times/

Silverman, P. (1997). A pragmatic approach to the inference of animal mind. In R. Mitchell, N. Thompson & H. Miles (eds.) *Anthropomorphism, anecdotes and animals* (pp. 170–188). State University of New York Press.

Sołtysiak, T. & Nogalski, Z. (2010). The effects of social hierarchy in a dairy cattle herd on milk yield. *Polish Journal of Natural Sciences*, 25(1), 22–30. doi: 10.2478/v10020-010-0002-1

Stevenson, B.E. (2015). *What is slavery?* Cambridge: Polity Press.

Stuart, D, Schewe, R.L. & Gunderson, R. (2012). Extending social theory to farm animals: addressing alienation in the dairy sector. *Sociologia Ruralis*, 53(2), 201–222. doi: 10.1111/soru.12005

Sunstien, C.R. (2004). Introduction: what are animal rights? In C.R. Sunstien & M.C. Nussbaum (eds.) *Animal rights: current debates and new directions* (pp. 3–18). Oxford University Press.

Talbot, M. (2013). Farm tourism in Wales, products and markets, resources and capabilities: the experiences of six farm tourism operators. *European Countryside*, 4, 275–294. doi: 10.2478/euco-2013-0018

Tuan, Y.F. (1984). *Dominance & affection: the making of pets*. New Haven: Yale University Press.

Tully, P.A.G. & Carr, N. (2019). The oppression of donkeys in seaside tourism. *International Journal of the Sociology of Leisure*. 3(1), 53–70. doi: 10.1007/s41978-019-00047-x

USDA (2018). United States Department of Agriculture Livestock slaughter. Retrieved from: https://www.nass.usda.gov/Publications/Todays_Reports/reports/lstk1018.pdf

Walters, T. (2016). Using thematic analysis in tourism research. *Tourism Analysis*, 21(1), 107–116. doi: 10.3727/108354216X14537459509017

Ward, P., Clark, T., Zabriskie, R. & Morris, T. (2014). Paper/pencil versus online data collection. *Journal of Leisure Research*. 46(1), 84–105. doi: 10.1080/00222216.2014.11950314

Wilkie, R. (2010). *Livestock/deadstock*. Philadelphia: Temple University Press.

Yarwood, R. (2005). Beyond the rural idyll: images, countryside change and geography. *Geography*, 90(1), 19–32.

Young, J. & Carr, N. (2018). Introduction. In J. Young & N. Carr (eds.) *Domestic animals, humans, and leisure: rights, welfare, and wellbeing* (pp. 1–12). Abingdon: Routledge.

Yudina, O. & Grimwood, B.S.R. (2016). Situating the wildlife spectacle: ecofeminism, representation, and polar bear tourism. *Journal of Sustainable Tourism*, 24(5), 715–734. doi: 10.1080/09669582.2015.1083996

Part 3: **Value-added work**

José-Carlos García-Rosell, Linda Tallberg

7 Animals as tourism stakeholders: Huskies, reindeer, and horses working in Lapland

Introduction

Animals in tourism engage in many forms of labor, from pulling strength to speed and riding, among others (Fennell, 2012). Animal bodies play an important role as a source of power or comfort and curiosity for tourists watching or touching them (Coulter, 2016). Moreover, it is through the emotional and embodied engagement between animals and humans (tourists and guides) that animal-based tourism experiences are co-created (Bertella, 2014; Haanpää & García-Rosell, 2020). Although animals do not receive any direct financial compensation for their work, their human owners provide for their physiological needs (e.g. food, water, and shelter) with part of the money paid by tourists. Through their work and symbolic value, animals generate significant economic benefits for both their human owners and the tourism destinations where their labor is performed. Following Coulter's (2016) thoughts on interspecies solidarity, we argue that animals are not only tourism workers, but are also tourism stakeholders. We use an ethics of care framework (Connolly & Cullen, 2018; Wicks, Gilbert, & Freeman, 1994) to analyze animal tourism workers in Finnish Lapland, concluding that the human-animal relationship is largely based on contractual care by tourism entrepreneurs. As such, animal workers are seen in instrumental terms, but with concrete and distinct relations with their human owners. Customers, on the other hand, seem to view animal workers as having intrinsic value. Hence, we argue that animals become tourism stakeholders within this context through their close relationship to the traditional human stakeholder groups of customers and owners.

Considering the important role of animal labor in tourism and other organizational contexts, there is value in developing new ways of understanding stakeholders and their relationships. The tourism industry, with its complex and often messy entanglements between human and animal experiential encounters, offers a fruitful context to critically evaluate human-animal relations (e.g. Danby, Dashper, & Finkel, 2019; Haanpää et al., 2019) and, as in this chapter, the notion of stakeholder status. Burton and Dunn (1996), along with Wicks, Freeman, and Gilbert (1994), first argued for a feminist framing in stakeholder theory in terms of establishing moral behavior, which has been somewhat overlooked in comparison to other normative moral justifications (such as rights/justice) within stakeholder theory. In this chapter, we continue the task of moving beyond the long-held deontological premises of stakeholder theory towards an understanding of stakeholders based on an ethics of care framework. Ethics of care is a core feminist framing of morality

https://doi.org/10.1515/9783110664058-007

based on relationships, connectedness, and interdependence. In an interview, Carol Gilligan, whose book *In a Different Voice* (1982) was foundational for a feminist ethics of care discourse, defines this as

> an ethic grounded in voice and relationships [. . .] the importance of everyone having a voice, being listened to carefully (in their own right and on their own terms) and heard with respect. An ethics of care directs our attention to the need for responsiveness in relationships (paying attention, listening, responding) and to the costs of losing connection with oneself or with others. Its logic is inductive, contextual, psychological, rather than deductive or mathematical. (Gilligan, 2011)

Following this, our aim is to examine how animals are tourism stakeholders by applying this feminist framework to animal labor. We note that the well-being of many animals in organizations can be seen as of key strategic importance, especially within animal tourism branding itself "responsible" or "ethical," and as such, the care of the animal workers can be seen as a vital aspect of the business strategy. However, we go further than merely suggesting that animals are of key strategic importance in these contexts. We suggest that animal stakeholder status is of inherent moral value and that the human-animal relationship already existing in practice supports this understanding.

Following the recent work by Connolly and Cullen (2018) applying an ethics of care framing of animals in organizations, we explore what this means for animal stakeholder status. Whereas a deontological approach to stakeholders seeks a fair solution between competing individual rights, an ethic of care sees the interests of carers, care-givers, and cared-for as importantly intertwined rather than simply competing (Held, 2006). In organizations where animals are workers, their interests are more than just aspects of being cared-for, as upholding the animal interests is vital to the success of the business. Drawing upon interviews conducted in Finnish Lapland with animal-based tourism entrepreneurs, and upon social media content produced by international tourists and local companies, we work towards a relational understanding of stakeholders, which gives primacy to the social bonds and emotions that exist between the human and animal tourism workers. We suggest that such a framing already exists in practice among some tourism entrepreneurs, who recognize that caring well for their animals affects their business in terms of receiving positive customer reviews. In this chapter, we explore what this means for stakeholder status beyond anthropocentric objectives.

Considering stakeholder theory

Who is a stakeholder?

There are various definitions of a stakeholder, but an overarching understanding of the concept is that a stakeholder is "any individual or group who can affect or is affected by the actions, decisions, policies, practices, or goals of the organization" (Carroll, 1996, p. 74). The term "stakeholder" is traditionally understood to include employees, customers, suppliers, investors, government officials, competitors, local community members, and activists, among others. Phillips, Freeman, and Wicks (2003, p. 480–481) define stakeholder theory "as a theory of organizational management and ethics" which is "distinct because it addresses morals and values explicitly as a central feature of managing organizations." This ethical and moral consideration of the interests of those affected by organizational decision-making is a core tenet throughout stakeholder theory, even though there are different scholarly opinions on who "legitimate" stakeholder status encompasses (see, for example, Phillips & Reichart, 2000). This core aspect to the theory is vital to recognize, as societal consideration increasingly acknowledges more-than-human actors (such as animals and nature) as beings and entities important not only to human survival and thriving, but with intrinsic value of their own.

Therefore, there has been much debate on whether non-human actors, such as the natural environment (which animals are traditionally merged into), should or can be considered as primordial stakeholders. Starik (1995), in his much-referenced article, made a convincing argument that the natural environment is a key stakeholder, separate from others such as suppliers, competitors, and stockholders, yet intertwined due to nature encompassing all other groups. We note that this ongoing scholarly discussion is multifaceted and based on the differing perspectives of rights, justice, and moral considerations an organization is obliged to consider. However, we suggest that in a time of impending climate change with systems of large-scale planetary disruption, it would be short-sighted (and not very ethical) to limit managerial decision-making only to human interests. In line with Gren and Huijbens (2014), we argue that the dire climate situation we face today (and in the future) is a result of overtly anthropocentric and individualist human enterprise based on outdated profit maximization ideals, something stakeholder theory, at its core, distances itself from as a moral theory (see, for example, Freeman 1984). Instead, we see that stakeholder theory presupposes the interconnectivity between sometimes diverse groups and individuals, a relational connection that feminist theory supports.

The normative justifications for stakeholder theory influence who to consider as a stakeholder. Phillips, Freeman, and Wicks (2003) conclude that there are eight different branches of the normative core of stakeholder theory. They split this core into (1) the common good, (2) feminist ethics, (3) risk, (4) integrative social contracts theory, (5) property rights, (6) Kantianism, (7) the doctrine of fair contracts, and (8) the principle of stakeholder fairness (for a detailed discussion of each, see Phillips, Freeman, & Wicks, 2003, p. 481). The widely held anthropocentric position of not considering animals (or nature) as stakeholders finds its origins in the latter, fairness-based principles, which have generally shaped understanding of the stakeholder notion. Such an approach does not necessitate aspects of caring or emotion, upon which we and many other feminists would argue relationships are built.

However, recent work has begun to consider animals as stakeholders due to increased societal concern for and understanding of animals in society. In *Animals and Business Ethics*, Smart (forthcoming) makes the case that animals should be treated as stakeholders due to them having moral status. He claims that as animals' interests are affected by organizational decisions and actions, we need to take such interests into consideration according to the core tenets of stakeholder theory (Smart, forthcoming). Following Carroll's (1996) definition at the start of this section – that it is those whose interests are affected by organizational decisions and actions who need to be considered as stakeholders – offers a logical rationale for going beyond previous business considerations and an anthropocentric focus to include animals.

An ethics of care perspective on stakeholder theory

In order to develop this consideration of animals as stakeholders, we use an ethics of care framework for working animals in tourism. Noting that animal tourism is a contested sector due to fluctuating levels of animal welfare within the global tourism industry, we suggest not only that animals benefit from being considered stakeholders, but that businesses reap benefits from the positive customer experiences that ethical interactions with animals provide in tourism destinations.

Using feminist moral ethics in stakeholder theory was first argued over 25 years ago (Burton & Dunn, 1996; Freeman & Gilbert, 1992; Wicks, Gilbert, & Freeman, 1994). As stakeholder theory is based on the morals and behaviors of management, it requires moral consideration beyond justice and rights (Burton & Dunn, 1996). Feminist moral ethics focus on caring, relationships, collaboration – many of the values which modern society is increasingly espousing, and even more so than when this framing was first suggested. This relational aspect of feminist ethics is of key importance. As Burton and Dunn (1996, p. 135) state:

> Feminist ontology of humans as essentially relational beings is tied to their epistemology. They believe that humans only know through relationships. If relationships are identifying characteristics of humans, then the abstract, universal, impartial, and rational standard of what is knowledge is inadequate in decision-making.

Emotion and affect are sources of knowledge in the feminist perspective, and elements of caring are highly relevant to moral behavior (Held, 2006). We note that applying an ethics of care framing to decision-making and stakeholders does not exclude aspects of rights, duties, and justice. Rather, this is a complementary layer to stakeholder theory. As Wicks (1996) points out, there are practical objections to using only an ethics of care framing to managerial decision-making, as in a practical sense there are no guidebooks or rules on managerial caring per se.

Applying an ethics of care framework to animals in organizations is not new. However, our goal in this chapter is to apply this framing to tourism studies as a unique context where animal workers are vital to the sector and where caring already occurs to some degree. In a review of the organizational studies field in relation to animals, Connolly and Cullen (2018) mapped publications from 1995 to 2015 on animals within business and management studies according to an ethics of care framing in four distinct caring categories: (1) there was "contractual care" between humans and animals; (2) there was a "no care environment," where care for animals was not considered, although they were present in the context; (3) there was a "care about" animals context or topic; and (4) there was a context where animals were "care[d] for" (p. 410). The articles revealed differences in seeing animals as providing either an instrumental value to the organization or having intrinsic value in themselves. Furthermore, the relationship with the animals was seen to either be "concrete," which was "characterised by a direct and personal interaction" with animals, or "abstract," where there was "an objective distance between human and animal" (Connolly & Cullen, 2018, p. 409). The authors found that those studies focusing on animals as instrumentally valuable in a concrete aspect applied a "contractual care" framing. These included farmers or ranchers who had a personal relationship with the animals, although these were seen in instrumental terms. Abstract notions of animals serving instrumental values were seen in the "no care environment," which included animals in entertainment, animals as research tools, and other forms of commodifying animals. On the other hand, animals with intrinsic value were seen in abstraction in articles focused around topics related to ethical consumption, animal advocacy, and sustainability in the "care about" category, while the fourth category, "care for," was focused on animal shelter workers and aspects related to companion animals who were seen to have intrinsic value and enjoyed a concrete human-animal relationship.

In the categorization (which included 185 peer-reviewed articles), 10 articles were focused on the tourism sector. Most of these (six articles out of 10) involved a "care about" relationship with animals. Within this, the subcategory of sustainability or conservation aspects in relation to animals was dominant (Higham & Shelton, 2011; Orams, 2002; Reynolds & Braithwaite, 2001). The second most dominant view from an ethics of care perspective was a "no care" relationship (three articles of ten), where animals were viewed as a marketing tool (Okello, Manka, & D'Amour, 2008), seen as a commodity (Galloway & Lopez, 1999), and seen as a source of disease (Lee & Chen, 2011). Only one article had "contractual caring" by farmers as its focus and, interestingly, none of the reviewed articles found a "care for" framing within tourism. Such a categorization is useful to apply to animal tourism workers, and we apply this ethics of care framing in the specific context of Finnish Lapland to better understand the situation in practice.

Animal workers in tourism: Husky, reindeer, and horse tours in the winter wonderland

Animal workers are vital to much of the tourism industry, which relies on encounters with nature and animals for unique tourism experiences in holiday destinations. Coulter (2016) creates a case for animals as workers in describing the entanglements that exist between humans and animals at work. She comments in her book *Animals, Work and the Promise of Interspecies Solidarity* (2016, p. 11) that "a large majority of labor researchers fail to see that humans are but one of many species in any given space and community, that many humans work with animals, that humans depend upon the broader ecological web for subsistence and survival, and that human, animal, and environmental well-being are inextricably connected." Coulter calls this framework "interspecies solidarity" to account for humane work which includes "jobs that are good for both people and animals" (p. 163). She concludes that work which exploits animals is often detrimental to humans and the environment, too (for example, slaughterhouse work, which impacts all organizational actors negatively). In this way, focusing on the well-being of humans and animals in solidarity benefits all involved, as the welfare of all actors is interlinked (see García-Rosell & Hancock, forthcoming). This inter-relational conceptualization links to our framing of animals as stakeholders in tourism, as the human-animal relationship is at the core of the context of the Finnish Lapland destination.

Finnish Lapland is the northernmost province of Finland and the European Union. With only 3.5% of Finland's population but about 30% of Finland's total area, Lapland is by far the least densely populated region in Finland, and a rapidly growing wilderness and nature-based tourism destination. The number of annual registered overnight stays in Lapland is 3 million (Regional Council of Lapland 2019). As a tourism destination, Lapland attracts tourists from around the world, particularly in the winter season between December and March. Although most tourists come from European countries, the number of Asian tourists has increased sharply in recent years. In Lapland, tourists engage in winter activities such as viewing the northern lights (or Aurora Borealis), driving snowmobiles, visiting Santa Claus, and riding reindeer sleighs and dog sleds. According to a recent study (García-Rosell & Äijälä, 2018), more than half of the tourists traveling to Finnish Lapland consider the winter landscape and animal-based activities the main reasons to visit this Nordic destination.

Animal-based tourism services, such as dog sledding and reindeer sleigh tours, are some of the most popular activities among tourists (García-Rosell & Äijälä, 2018). Indeed, these two activities form part of the ritualized pattern that international tourists follow when visiting Lapland. A recent study calculated that there are 4,000 dogs and 750 reindeer working in the tourism industry in Lapland, generating an annual turnover of approximately 15 million euros (García-Rosell & Äijälä, 2018).

However, this is an estimate, as their actual number may be higher, especially the number of dogs. There are also 150 horses working in tourism, but these tend to be more popular among the domestic Finnish visitors than international ones. Although the reindeer is an icon of Lapland and an essential element of the Christmas narratives, it is husky activities that are growing in popularity. In 2017, for the first time, husky safaris surpassed snowmobile safaris as the number one tourism activity in Lapland. Similar to human employees, the working and resting hours of these animal workers are carefully planned by their owners, and most of their work is done during the winter period (Ojuva, 2018).

Huskies in Lapland live in outdoor dog kennels with populations that can vary from a dozen to 500 dogs. The number of all-year-round kennels in the Finnish Lapland area is close to 50. In addition, mushers (dog handlers) come with their dogs from other parts of Finland and Europe to work in the industry during the peak season (December–February). The number of seasonal mushers is difficult to determine due to a lack of official statistics. Most of the sled dogs in Lapland are Alaskan huskies, which are a mix of different northern breeds chosen particularly for their pulling skills. There are also Siberian huskies, which is the breed commonly depicted in the commercial marketing and internet material for tours. Husky safaris can range from short (0.5–2 km) to medium-length (10–40 km) and multi-day rides (two to eight days). The sleds are pulled by four to six dogs depending on the size and weight of the musher and passenger(s). Almost all of the safaris are organized in the same way. Firstly, independent of the length of the safari, tourists receive driving and safety instructions before entering the kennel. Secondly, driving instructions are provided by a guide demonstrating how to handle the sled and the team of dogs. Thirdly, the tourists are guided towards the sleds to get ready for the start of the safari (Figure 7.1). Depending on the type of safari, during the ride there are different stops where the driver can swap places with the passengers. In this way, all the adult tourists have an opportunity to drive the sled. When the safari is over, tourists can take a tour of the kennels and learn more about and cuddle the dogs.

For many tourists, just seeing reindeer along the road or while walking through the forest in the snow is part of the unique Lapland tourism experience. Like husky tours, reindeer sledging is very popular with tourists of all ages throughout the winter season, and in the Christmas season in particular (see García-Rosell & Hancock, forthcoming). Reindeer work may take different forms (see Hoarau-Heemstra, 2018; Nieminen, 2014). Some reindeer are enclosed within the boundaries of the reindeer farms, where tourists can approach and feed them, as well as photograph themselves with the exotic and mythical animals. Other reindeer work as draft animals, pulling sleighs in activities. In the tourism industry, reindeer perform the roles of both workers and objects of the tourist gaze (Urry, 1990). The reindeer working as draft animals are castrated male reindeer who have undergone a long period of training, lasting between four and five years. In comparison to dog sledding, tourists do not need to handle the sleigh. These tours may take place in an enclosure where the reindeer

Figure 7.1: Preparing for the start of the sled dog safari.
Source: Author.

follow a predetermined track (Figure 7.2). In case of longer tours, a guide will drive in front of all sleighs, leading and setting the pace of the tour. Reindeer tours may last from half an hour to a couple of hours. Female reindeer are not used as draft animals due to the gestation period that takes place during the winter season. Reindeer only pull sleighs during the winter season from December to April, and most of the tourism reindeer roam free in the pastures and forest of Lapland in the summer. A small number of reindeer stay at the reindeer farm during the summer months for tourist "meet and greets." As part of the reindeer herding annual cycle, female and unskilled male reindeer are slaughtered for human consumption as in poronkäristys (traditional sautéed reindeer) and their bones, antlers, hides, and other body parts used to produce traditional tourist souvenirs, (see García-Rosell & Hancock, forthcoming; Hoarau-Heemstra, 2018).

Although horses are less popular among international tourists, there are several horse stables offering horseback riding in Lapland. Some Finnish tourists prefer to experience horse-riding in Lapland, rather than the husky or reindeer tours which are more popular with international visitors. These tourists vary in their experience; some domestic tourists are advanced riders, while others may have limited riding

Figure 7.2: Reindeer sleigh rides at reindeer farm.
Source: Marko Junttila. Used with permission.

experience. Safety and winter wear is often made available to visitors, and they will have at least one experienced guide during the riding tour. Most horses used for horseback riding in Lapland are Finnhorses or Icelandic horses. The Finnhorse is a domestic horse of Finland and serves as the country's official national horse breed. Icelandic horses are also popular in tourism activities in Lapland. Both horse breeds are capable and reliable mounts, well adapted to the harsh climate and terrain of the region. As both the Finnhorses and Icelandic horses grow a thick winter coat, they can tolerate surprisingly low temperatures. The horses usually work in trail riding of different lengths from one hour to a half day or entire day. Some companies also offer longer trail rides which include overnight stays in cabins along the trails. The horse stables in Lapland have between five and 20 horses. Although stalls are common practice, the number of loose stables is growing in Lapland (Figure 7.3). A loose stable allows horses to live in a herd and move freely both outdoors and indoors. As with the huskies and reindeer, the work shifts of horses are monitored and the individual character and strength of horses is considered in shift planning by the human caretakers (Kähkönen, 2018).

Figure 7.3: Loose stable at horse farm.
Source: Author.

Animal stakeholdership: Social bounds and emotions

In this chapter, we aim to illustrate animals becoming tourism stakeholders by applying a feminist normative core of stakeholder theory (Burton & Dunn, 1996; Wicks, Gilbert, & Freeman, 1994), specifically that of ethics of care. The excerpts used in this chapter come from interviews with animal-based tourism entrepreneurs and from social media content generated by Lapland tourists and service providers. The data were gathered within the scope of two parallel two-year EU-funded research and development projects focusing on animal welfare as part of responsible tourism in Finnish Lapland. Data collection was conducted by four researchers, one project worker, and five Bachelor's/Master's students. The first author of this chapter was one of the researchers and the leader of the team. During the early stage of the two projects, the team collected documentary material about the topic in general and in relation to Finland and Lapland in particular, implemented a survey with tourists, conducted participant/non-participant observations, and attended several meetings in the premises of approximately 20 different animal-based tourism companies. Based on this preliminary work, companies that showed a high level of commitment towards animal welfare were selected to be interviewed. The social media data were collected in the last stage of the projects to complement the data gathered by examining online discussions related to animal welfare in Lapland.

A total of 11 semi-structured interviews with the owners of husky kennels and reindeer and horse farms operating in Finnish Lapland were conducted by one researcher and the project worker, occasionally accompanied by one or two Bachelor's students, between March 2016 and May 2017. The interviews took place in the premises of the companies and concentrated particularly on the treatment and welfare of their animals in relation to their business operations, customers, and communication strategies. The interviews lasted approximately 60–90 minutes; they were audio recorded and then carefully transcribed. The user-generated social media content included publicly available comments, reviews, and discussions created during 2016, 2017, and January 2018. The content used relates to 45 Lappish tourism companies and consists of 304 reviews systematically collected from Facebook, Instagram, TripAdvisor, and YouTube during January and February 2018 by one Master's student under the supervision of the first author of the chapter (Klos, García-Rosell, & Haanpää, 2018). The companies were selected based on their social media presence from a list of 158 animal-based tourism service firms operating in Lapland (Bohn, García-Rosell, & Äijälä, 2018). Furthermore, only companies which offer animal-based activities involving sled dogs, reindeer, and horses were considered in the study. Destination management companies that sell animal-based activities were also included in the study due to their significant presence on social media.

The data were thematically analyzed relying on content analysis (Bengtsson, 2016). The interview data were divided according to five main themes: animal welfare, working relationships, customer relationships, communication, and safety. The social media content categories were established inductively in relation to the collected data. By carefully reading through the social media data, we were able to identify two main themes: animals considered in the comments and animals omitted in the comments. In this chapter, we focus on the theme of animals considered in the comments, which was divided in four sub-themes: animal welfare, animal appearance, animal behavior, and animals as educational resource. After discussing and reflecting on the data collaboratively, we identified passages reflecting the care perspective on human-animal stakeholder relationships. Next, we present the results of the analysis by first discussing animal stakeholder status being embedded in the social bonds with their owners, and then discussing the emotions and social bonds evolving through customer interactions.

Animal stakeholder status embedded in social bonds with owners

Using an ethics of care perspective on this data draws attention to relationships as morally worthy and of value in human-animal interactions. Animal-based tourism entrepreneurs tend to stress a caring relationship when talking about their animals. As one reindeer tourism entrepreneur points out:

> These reindeer are our colleagues and, like, we know every reindeer personally [. . .] and [their] characteristics, in a way that it is our partner and not [just] a tool.
>
> (Excerpt 1, Interview Reindeer Farm)

This focus on it being a collaborative effort, a form of partnership, fits well with feminist moral theory, as well as the interspecies solidarity of Coulter (2016). This is particularly relevant when further describing the relationship between animal-based tourism entrepreneurs and their animals. As Excerpt 2 shows, entrepreneurs tend to give preference to those animals with whom they have a close relationship. This follows the ethics of care framing, as those we relate to are shown higher concern and care than those further away from us (similar to social bonds) (Held, 2006). For example, as one of the reindeer tourism entrepreneurs expresses:

> When a reindeer becomes too old, we might send it to the slaughterhouse, but it can also happen that we put it down ourselves and take it out into the forest [. . .] but I wouldn't feel like selling an old sledge reindeer as meat, because it has been working for so many years and it has been an everyday workmate [. . .] so it [what happens to aging animal workers] depends a bit on the situation. (Excerpt 2, Interview Reindeer Farm)

This relationship may not necessarily eliminate harm, but it may give certain "privileges" to animals who share a special bond with the owners of the company. In terms of relating both interview excerpts to the ethics of care- categorization of Connolly and Cullen (2018), it seems that the reindeers and their owners have a contractual care relationship here. As such, it seems that some of the animal workers in this specific tourism destination might benefit from a closer relationship than in other contexts: although these are special connections with certain individual animals or groups of animals, it does not give them intrinsic value as the "caring for" categorization would. For example, the quote above reveals that the entrepreneur would not be able to sell a "workmate" as meat once he is "too old" to work by pulling a sledge, but may instead be able to kill it themselves and "take it out into the forest." The reindeer is seen in instrumental terms as he is killed once "spent," suggesting the commodification of the animal's utility over an inherent intrinsic value (Hughes, 2001), which the animal shelter and companion animal category in Connolly and Cullen's (2018) categorization focused on. Linguistics also matter in revealing this instrumental relationship. The entrepreneur refers to the reindeer as "it" rather than he, suggesting a distancing and objectification of the animal apart from human workers.

The emphasis on collaboration and cooperation – characteristic of the feminist perspective in stakeholder thinking, as described by Wicks, Gilbert, and Freeman (1994) – is evident in the ways in which the human-animal relationships are developed in the animal-based tourism companies. Some describe that there is a mutual understanding between the humans and the animals that makes things work, as one of the horse entrepreneurs explain:

> As long as the horse is lively and cooperative, when you tell it "let's go for a ride" and it's willing to come, you know it's fine. You know it when a 600 kg creature isn't willing to come. And if it has an appetite for its food, I figure the animal is doing well. And when you know approximately its limits, you make sure not to bring the horse close to them. We don't want that to happen, as everything would stop working. (Excerpt 3, Interview Horse Farm)

The ethics of care framework can easily be applied to the context, although to different degrees in terms of its practical application. It seems that caring, relationship-building and cooperation serve as mechanisms for the humans and animals to learn from each other, build trust, and find ways of working together in order to deliver the service – the animal tour experience. This aspect of collaboration in the feminist framing seems to transcend specific species in the Lapland context, applying to horses as well as dogs as animal workers.

> You've got to think about it in the way that you treat each dog as if it was your own. Because when it's your own dog, it likes you and doesn't run away with its tail between its legs but comes to your side. Only then is it easy to deal with them, when they like to be with you and only then do they run the best and work well [. . .] The same with the horses. It's not going to work by beating it with a stick. On the contrary, you have to work on building trust in the horse so that when it comes to a difficult situation, the horse will trust you and follow your instructions. (Excerpt 4, Interview Kennel and Horse Farm)

Relationships built in an animal-based tourism context seem not only to contribute to facilitating close proximity between human and animal workers, but also to cultivating the virtues of care, respect and self-mastery (see Anthony, 2012). From this perspective, an ethics of care provides an opportunity for human workers to remain close to the outcome of their decisions and those animals that will be directly affected by them (Connolly & Cullen, 2018). This is reflected in the following quote from one of the sled dog entrepreneurs:

> My husband is very good at spotting dogs that aren't performing well in the safari. Sometimes you need to change the dogs during the trip. It's very important that you know how to do it. If one of our dogs doesn't tolerate a customer, for example, because they brake wrong or ride too fast downhill, my husband stops the safari and replaces the dog with one that's better at keeping up with that type of customer [. . .] So, we are pretty good at noticing these individual aspects. (Excerpt 5, Interview Kennel and Horse Farm)

This quote shows the importance of reading the dogs' behavior and knowing their different temperaments with difficult customers. Without this mutual respect and understanding between the sled dog entrepreneur and their dogs, the dogs would suffer at the hands of tourists who, more often than not, are unskilled in driving sleds and handling dogs. In this way, it is beneficial both for the dogs, who are replaced if not "performing well," and for the business owner, who inevitably will gain better customer reviews from positive sled experiences. From the dogs' perspective, dealing with such a variety of customers, it is essential that the owners pay attention to reading customers and their behaviors. One could ask whether it is appropriate for all customers to be given the opportunity to handle a dog sled, and at what point the owners take over if the customer is not able to handle themselves appropriately; however, such aspects were not available in our data. But seeing the sled dogs as key stakeholders, with their own limitations and strengths, may lead to overall better experiences for the humans and animals involved, in addition to positively affecting the business outcomes.

Animal stakeholder status embedded in customer social bonds and emotions

Another human-animal relationship that exists in this context is the social bond between animal and customer. The customer comes to experience an animal adventure at the tourism destination and therefore leaves with an emotive human-animal experience. Customers are already considered key stakeholders in most service businesses, and customer satisfaction is usually upheld as a barometer for good business dealings. Within the tourism industry, online tourist customer reviews play a key role in generating new business, with social media comments used to tell experiential stories as part of tourism companies' communication strategies. We use quotes from both company-generated content and customer reviews to analyze the ethics of care framing for animal stakeholder status based on the social bond with the customer. We contend that the social bond with customers (a key traditional stakeholder group) and the emotional aspects of human-animal interactions are what assigns stakeholder status to animals. Feminist moral theory acknowledges emotions as key areas of knowledge (Burton & Dunn, 1996), and here we see the use of such aspects for commercial benefits, as well as underpinning the experience. Emotions are key marketing tools, and positive aspects regarding the animals and their welfare are important for more business.

> The staff are very helpful and accommodating and they really care for their dogs. I was a bit sceptical of these kinds of things [animals' quality of life], but the company really opened my eyes to the world of sled dogs. These beautiful dogs love what they do and they are properly cared for and very much loved. (Social Media Content, Customer A)

> Honestly, the thing that I loved the most about this experience was how much the business revolves around the dogs. The owner has a special formula for their food, supplemented with a variety of meats. The staff members know the names of all 95 dogs, and spend 2.5 years working with

> them before they ever pull a sled, and their work schedule is determined by the dog's physicality [. . .] dogs who need more down time get it, while dogs that love to run get sent out more. They treat their dogs like family, and it was really heart-warming to get to see it.
> (Social Media Content, Customer B)

> I did a lot of research into which husky safari to choose and wanted to book with one which had good ethics and clearly cared for the dogs. From the moment I started corresponding with [the owner of the kennel], I knew that this was important to him and the farm also! The dogs clearly loved working and enjoyed some cuddles afterwards. (Social Media Content, Customer C)

All three accounts demonstrate the importance of the dogs' well-being. They show how some animal-based tourism customers to this particular destination look for, research, and evaluate whether the animals they encounter when buying the animal experiences are well cared for and "loved" as "family." Although it would be interesting to know more about what "good ethics" means for Customer C and whether all reviews (and not only positive ones) are available to customers, the comments show the importance of "good" animal care from the customers' perspective. In this way, from an ethics of care perspective, the customer's social bond to the dog becomes more of a "care about" framing rather than the overt "contractual care" seen in the owner-dog relationship earlier.

However, in the company-generated content, there were also aspects of owners "caring for" their dogs. The following quote is from a company Facebook page demonstrating a "care for" relationship with an ex-sled dog by adopting him to a home rather than euthanizing him, which is a normal and fully legal procedure in Finland and most countries. When a sled dog is unable to work due to behavioral issues (being too timid or aggressive) or physiological issues (being too weak to pull the sled or too old), the kennel can legally euthanize the dog. One of the most controversial aspects of dog sledding is that euthanasia is used to a large degree to manage dogs who are not good workers. As shown in the previous section, many view the animals from a "contractual caring" perspective, which this management style would support. Creating systems of adoption or other alternatives to euthanasia would require owners to shift more towards the "caring for" perspective. The following sled entrepreneur wants to show that responsible animal tourism is possible through applying a "care for" aspect, intrinsically valuing the dogs for themselves, rather than purely for the value that the business generates from them. As such, they use adoption as an alternative to euthanizing the dogs.

> Miyagi is super cuddly and loves attention and lots of food. Miyagi was castrated a few years ago because he could be a bit aggressive towards other dogs. He is much better now, but sometimes still snaps out of excitement, which can irritate his running partner and lead to fights, but he doesn't really mean any harm. He simply lacks a bit of social competence. He is a dog that it's hard not to like, and he certainly loves people. This big-eared loveable goofball has been a favourite of many guides over the years, but this year he finally hit the jackpot when he met Alva who came from Sweden to work as a guide at the beginning of March. Around this point in the season, Miyagi (who was never really a very good sled dog) decided he would

> rather be on a couch than pulling a sled, and was fast-tracked onto our adoption list. Alva had already fallen in love with his goofy personality, and decided to take him on. In a few days, Alva and Miyagi will begin their journey to their new home outside Stockholm where they will live and work in a horse stable. (Social Media Content, Company H)

Such an account clearly shows a different type of agency of the dog who is given intrinsic value beyond his ability to work (see Hughes, 2001).

Conclusions

In this chapter, we suggest that applying an ethics of care framing to animal workers in tourism shows that there are already degrees of animal stakeholder status in the industry. Relying on empirical data from animal-based tourism firms operating in Finnish Lapland, we demonstrate that what makes a husky, reindeer, or horse a tourism stakeholder in the first place are the social bonds and emotions that they share with human stakeholders, such as their owners and customers. These findings are portrayed below in Figure 7.4, adapted from Connolly and Cullen (2018). We find that the human-animal social bond with owners is predominantly based on what Connolly and Cullen label "contractual care," whereby the human-animal relationship is distinct but often instrumental.

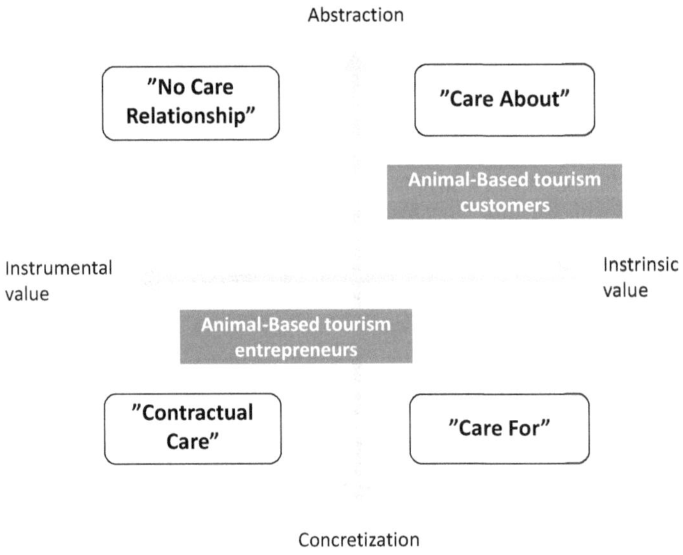

Figure 7.4: Ethics of care framework applied to animal-based tourism in Lapland.
Source: Adapted from Connolly and Cullen (2018, p. 410)

On the other hand, the social bonds and emotions of customers' reviews are based on what Connolly and Cullen label "care about," where the animal has greater intrinsic value, yet there is still an abstract or indirect relationship with the animals. We found that there are some animal tourism entrepreneurs who are moving towards Connolly and Cullen's "care for" categorization of the human-animal relationship through using adoption programs for dogs in retirement or animal workers with poor performance. We find this to be a positive move, whereby the animal workers truly gain agency and value beyond their direct commercial value. Nevertheless, animal work is still not covered by labor legislation, which is exclusively reserved for human workers. Indeed, the question remains: should the animal workers not have some of the benefits human workers enjoy beyond their working role – for example, pension plans, sickness benefits, and value beyond utility? These are some of the ethical considerations that need to be addressed if we truly wish to employ animal workers in our businesses. Moving towards a "care for" relationship, where each animal's unique individuality and right to life beyond commercial gains are ensured, would surely lead to more authentic and responsible human-animal interactions in tourism (see Fennell, 2014). This could lead to further interspecies solidarity and humane work (Coulter 2016), which could create new ways of relating to and with animals at work.

References

Anthony, R. (2012). Building a Sustainable Future for Animal Agriculture: An Environmental Virtue Ethics of Care Approach within the Philosophy of Technology. *Journal of Agricultural and Environmental Ethics 25*, 123–144.

Bengtsson, M. (2016). How to Plan and Perform a Qualitative Study using Content Analysis. *NursingPlus Open* 2, 8–14.

Bertella, G. (2014). The Co-creation of Animal-based Tourism Experience. *Tourism Recreation Research* 39(1), 115–125.

Bohn, D., García-Rosell, J. C., & Äijälä, M. (2018). *Animal-based Tourism Services in Lapland. University of Lapland, Multidimensional Tourism Institute*. Available at www.matkailuelaimet.fi.

Burton, B. K. & C. P. Dunn. (1996). Feminist Ethics as Moral Grounding for Stakeholder Theory. *Business Ethics Quarterly* 6(2),133–147.

Carroll, A. B. (1996). *Business and Society: Ethics and Stakeholder Management*. 3rd edition Cincinnati, OH: Southwestern.

Connolly, L., and J. G. Cullen. (2018). "Animals and Organisations: An Ethic of Care Framework." *Organization & Environment* 31(4), 406–424.

Coulter, K. (2016). *Animals, Work, and the Promise of Interspecies Solidarity*. New York: Palgrave Macmillan.

Danby, P., K. Dashper, & R. Finkel. (2019). "Multispecies Leisure: Human-Animal Interactions in Leisure Landscapes." *Leisure Studies* 38(3), 291–302.

Fennell, D. (2012). *Tourism and Animal Ethics*. Abingdon: Routledge.

Fennell, D. (2014). "Exploring the Boundaries of a New Moral Order for Tourism's Global Code of Ethics: An Opinion Piece on the Position of Animals in the Tourism Industry." *Journal of Sustainable Tourism* 22(7), 983–996.

Freeman, E. (1984). *Strategic Management: A Stakeholder Approach*. Boston, MA: Pitman.

Freeman, R. E. & Daniel R. Gilbert. (1992). "Business Ethics and Society: A Critical Agenda." *Business & Society* 31, 9–17.

Galloway, G. & K. Lopez. (1999). "Sensation Seeking and Attitudes to Aspects of National Parks: A Preliminary Empirical Investigation." *Tourism Management* 20, 665–671.

García-Rosell, J. C. & M. Äijälä. (2018). "Animal-based Tourism in Lapland." In *Animal Welfare in Tourism Services: Examples and Practical Tips for the Well-being of Animals used for Tourism in Lapland*, edited by J. Ojuva, 10–24. Rovaniemi: Lapland University of Applied Sciences.

García-Rosell, J. C. & P. Hancock. (Forthcoming). "Working Animals, Ethics and Critical Theory." In *Animals and Business Ethics*, edited by Natalie Thomas. Cham: Springer.

Gilligan, C. (1982). *In a Different Voice: Psychological Theory and Women's Development*. Boston, MA: Harvard University Press.

Gilligan, C. (2011), July 16. "Care Ethicist Network Interview with Carol Gilligan." https://ethicsofcare.org/carol-gilligan/

Gren, M. & E. Huijbens. (2014). "Tourism and the Anthropocene." *Scandinavian Journal of Hospitality and Tourism* 14(1), 6–22.

Haanpää, M. & J. C. García-Rosell. (2020). "Understanding Performativity and Embodied Tourism Experiences in Animal-Based Tourism in the Arctic." In *The Routledge Handbook of Tourism Experience Management and Marketing*, edited by S. K. Dixit, 229–237. London: Routledge.

Haanpää, M., T. Salmela, J. C. García-Rosell, & M. Äijälä. (2019). "The Disruptive 'Other'? Exploring Human-Animal Relations in Tourism through Videography." *Tourism Geographies*. https://doi.org/10.1080/14616688.2019.1666158

Held, V. (2006). "The Ethics of Care." In *The Oxford Handbook of Ethical Theory*, edited by D. Copp, 537–566. Oxford: Oxford University Press.

Higham, J. E. S. & E. J. Shelton, (2011). "Tourism and Wildlife Habituation: Reduced Population Fitness or Cessation of Impact?" *Tourism Management* 32, 1290–1298.

Hoarau-Heemstra, H. (2018). "A Life Worth Living: Reindeer in Nordic Tourism Experiences." In *Animals, Food and Tourism*, edited by C. Kline, 113–128. London: Routledge.

Hughes, P. (2001). "Animals, values and tourism – structural shifts in UK dolphin tourism provision." *Tourism Management* 22, 321–329.

Kähkönen, O. (2018). "Animal-based Tourism in Lapland." In *Animal Welfare in Tourism Services: Examples and Practical Tips for the Well-being of Animals used for Tourism in Lapland*, edited by J. Ojuva, 10–24. Rovaniemi: Lapland University of Applied Sciences.

Klos, D., J. C. García-Rosell, & M. Haanpää. (2018). "Analysis of User-generated Social Media Content related to Animal-based Tourism Activities in Lapland." Accessed May 20, 2020, www.animaltourismfinland.com

Lee, C., & C. Chen. (2011). "The Reaction of Elderly Asian Tourists to Avian Influenza and SARS." *Tourism Management* 32, 1421–1422.

Nieminen, M. (2014). *Poro-Reindeer*. Helsinki: Books on Demand.

Ojuva, J. (2018). *Animal Welfare in Tourism Services: Examples and Practical Tips for the Well-being of Animals used for Tourism in Lapland*. Rovaniemi: Lapland University of Applied Sciences.

Okello, M. M., S. G. Manka, & D. E. D 'Amour, (2008). "The Relative Importance of Large Mammal Species for Tourism in Amboseli National Park, Kenya." *Tourism Management* 29, 751–760.

Orams, M. B. (2002). "Feeding Wildlife as a Tourism attraction: A Review of Issues and Impacts." *Tourism Management* 23, 281–293.

Phillips, R. A. & J. Reichart. (2000). "The Environment as a Stakeholder? A Fairness-based Approach." *Journal of Business Ethics* 23, 185–197.

Phillips, R. A., Freeman, R. E., and Wicks, A. C. (2003). What Stakeholder Theory Is Not. *Business Ethics Quarterly* 13(4), 479–502.

Regional Council of Lapland. (2019). *Lapin matkailustrategia 2020–2023* [Lapland Tourism Strategy 2020–2023]. Lapin liiton julkaisuja Serie A, No. 54. Rovaniemi: Regional Council of Lapland.
Reynolds, P. C., & D. Braithwaite. (2001). "Towards a Conceptual Framework for Wildlife Tourism." *Tourism Management* 22, 31–42.
Smart, J. (Forthcoming). "Animals as Stakeholders." In *Animals and Business Ethics*, edited by Natalie Thomas. Cham: Springer, accessed May 1, 2020, https://philarchive.org/archive/SMAAAS-2
Starik, M. (1995). "Should Trees have Managerial Standing? Toward Stakeholder Status for Non-human Nature." *Journal of Business Ethics* 14(3), 207–217.
Urry, J. (1990). *The Tourist Gaze*. London: Sage.
Wicks, A. C. (1996). "Overcoming the Separation Thesis: The Need for a Reconsideration of Business and Society Research." *Business & Society* 35(1), 89–118.
Wicks, A. C., D. R. Gilbert, & E. Freeman. (1994). "A Feminist Reinterpretation of the Stakeholder Concept." *Business Ethics Quarterly* 4(4), 475–497.

Hindertje Hoarau-Heemstra, Nadezda Nazarova

8 Distributed leadership in tourism experiences: Russian sled dogs and Icelandic horses leading the way

Some people talk to animals. Not many listen though. That's the problem.
— A.A. Milne, Winnie-the-Pooh (1926)

Introduction

Animals play an important role in the creation of nature-based tourism experiences, such as horseback riding and dog sledding. Animals are key front-line players in the tourism industry, serving guests during their journey and sometimes even acting as leaders. For example, horses know where to go, even with inexperienced riders, and experienced horses can handle humans who are not familiar with riding at all. Similarly, sled dogs can survive in harsh Arctic environments without human assistance. They know the weather and topography and have experience in finding their way in difficult weather conditions and terrains. Based on and as a result of such natural animal abilities (Gibson, 1979), humans involved in the experience (as guides or tourists) sometimes have to hand over their agency to the animals, trust the animals, and follow their animal leaders to have an enjoyable and safe tourism experience.

However, this animal-based leadership in the context of the tourism experience has not yet been properly explored, calling forth a number of questions. For example, in regard to listening to animals in this context – how do we do that and why should we do it? To listen to animals, tourism actors need to understand the language of the animals they are working with. When animals are more understood in general, they can have more control over their environment and express more agency (Meijer, 2019). This is a topic of interest because the agency of animals in tourism is rarely discussed in the emerging stream of literature on animals in tourism. Animal welfare, the different uses of animals in tourism, and the effects of tourism on wildlife and domesticated animal species are addressed in the literature (see Fennell [2012] and Markwell [2015] for a thorough discussion on the relationships between animals and tourism). However, animals as agents of communication and leadership have not been considered.

Building on the post-heroic view of leadership, leadership activities are collective rather than individual (e.g., Grint, 2010; Harris, 2009; Spillane et al., 2001, etc.), this chapter is concerned with the distribution of leadership in animal-based tourism experiences, the way leadership roles can change throughout the tour, and the role of animals as leaders. We focus on the relationships between tour leaders,

https://doi.org/10.1515/9783110664058-008

animals, and tourists, how communication is handled within these relationships, and how leadership is distributed among the different actors throughout the tour. Communication is an essential element when sharing and distributing leadership. The research question we will therefore explore in this chapter is as follows: how is leadership distributed between humans and animals during a tourism experience? To do this, we present narratives from dogsledding in northern Russia and horseback riding tours in Iceland. The nature of the tourism experience in Arctic conditions creates the necessity to distribute leadership among all participants, which in turn could change the way we think about the role of animals in tourism.

Theoretical background

Leadership

Among social animals, leadership and followership emerge because there is a need for coordination between individuals. Across species, individuals are more likely to emerge as leaders if they have particular morphological, physiological, or behavioral traits that increases their propensity to act first in problems that require coordination to solve (King et al., 2009). Traits such as motivation, temperament, dominance, and knowledge influence the way leaders arise and are accepted by a group. The traditional view on leadership is based on the assumptions that people's powerlessness, lack of personal vision, and inability to master the forces of change are deficits that can be remedied only by a few great leaders (Senge, 1990, p. 340). Hence, traditionally, leadership has been formulated as a top-down phenomenon that arises from an individual, heroic leader.

Biological, social, and psychological approaches to leadership have much in common within both the human and animal contexts; one key to the emergence of leadership and followership is the need to coordinate (King et al., 2009). In the context of human organizations, leadership is the process of motivating, influencing, and directing others in the organization to work productively in pursuit of organizational goals. In animal-based studies, the concept of leadership is frequently used to account for the coordination of group movements by single individuals; one or a few steer the behavior of many. Leaders are traditionally described as the animals moving in the front position during travel, the first individuals to depart, or any animal that manages to recruit partners. All these definitions in mammals imply that one individual is consistently recognized as the group leader over time. In the human-leadership literature, this corresponds with the traditional view of leadership. When leading, an individual articulates a clear vision and energizes and enables organizational members to understand their part in achieving collective goals. Leadership involves leaders using their power, personality, influence, knowledge, persuasion, and communication skills to coordinate people and groups so that their activities and efforts are harmonious.

This understanding of leadership has been theorized as a trait-based approach and has been motivated by the search for specific key personal characteristics or traits that can precede leadership success, despite time and situational contexts (Zaccaro, 2007). According to this theoretical perspective, leaders are born into the role. A noteworthy critique of trait theory focuses on the fact that traits are circumstance-specific and often change over time (McGehee et al., 2015); in other words, they can also be developed and learned. To expand the conceptualization of leadership and include the role of followers, the concept of transactional leadership was developed. Transactional leadership is built upon the exchange of something valuable between a leader and their followers in exchange for increased productivity (Burns, 1978).

In some contexts, followers accept the decisions of a specific individual on a regular basis (despotism), while in others, almost any individual can, in principle, elicit followership (democracy). In each case, to avoid conflict as well as to achieve coordinated group action, individuals must agree with leaders on a particular course of action. There is considerable debate as to whether human groups are essentially democratic and egalitarian or despotic and hierarchical given that the former is more common among hunter-gatherer societies, whereas the latter is evident throughout recent history (King et al., 2009).

Cullen-Lester and Yammarino (2014, p. 1) have seen the transition from orthodox and "heroic" leadership towards collective forms of leadership as "a paradigm shift" within the field of leadership. This shift is due to the recognition that "teams, organizations, coalitions, communities, networks, systems, and other collectives carry out leadership functions through a collective social process" (Cullen & Yammarino, 2014, p. 1). Kirk and Shutte (2004) suggest that while there are many ways to lead, leadership is the process of harnessing a person's strengths to fulfill a role that is valuable for the community. In the tourism experience context, leadership involves the identification, acquisition, allocation, coordination, and use of the social, material, and cultural resources necessary to establish the conditions that allow the creation of a valuable tourism experience (McGehee, Knollenberg, & Komorowski. et al., 2015). This understanding supports a transformative perspective of leadership, defining it as the "ability to empower others" with the purpose of bringing about a major change in the form, nature, and function of some phenomenon (Spillane et al., 2001).

Additionally, in a rapidly changing and complex world, leaders might not have sufficient and relevant information to make highly effective decisions by themselves (Fitzsimons, Turnbull, & Denyer, 2011). As a result, there has been a growing interest within management and organization studies about alternative models of leadership in which leadership is not limited to the formally appointed leader. This has led to the emergence of so-called "distributed leadership" (or dispersed, devolved, democratic, shared, collaborative, collective, cooperative, concurrent, coordinated, or relational leadership, or co-leadership), which highlights leadership as an emergent property in a group of interacting individuals (Fitzsimons, Turnbull, & Denyer, 2011).

The distributed leadership perspective recognizes that there can be multiple leaders (Harris & Spillane, 2008) and that leadership activities are widely shared within and between organizations (Harris, 2009). A distributed perspective on leadership also recognizes the work of all individuals who contribute to leadership practice, whether or not they are formally defined as leaders (Spillane, 2012). The research on distributed leadership draws on distributed cognition and social learning theory. It is anchored in the idea that the existing capacities of individuals can be enhanced through social interaction and connecting sources. Distributed cognition suggests that capacities are distributed throughout the social and material conditions of an organization and that they are fluid rather than fixed (Harris, 2009). Distributed leadership is primarily concerned with the co-performance of leadership and the reciprocal interdependencies that shape that practice (Spillane, 2012). As such, this perspective suggests openness in leadership boundaries, which raises the question of who is involved in leadership or who contributes to leadership. Distributed leadership entails the view that various types of expertise are distributed among many people, not concentrated in a few people, so that everyone plays a leadership role at different times and can make a significant contribution (Woods, 2004). As a result, the boundaries of leadership become more porous, encouraging many more members of the community to participate in leading, which expands the degree of knowledge available to the community. This, Gronn (2003) argues, creates leadership synergy and creative action.

As with mainstream leadership research, studies on tourism have evolved from the trait-based approach to a more transformational focus on empowering followers and fostering trust and respect among them (Zach, 2016). However, most studies do not call for the sharing of power between leaders and followers (McGehee et al., 2015). In particular, Valente, Dredge, and Lohmann (2014) have suggested that distributed leadership provides the opportunity for equitable power and decision-making among many participants in a tourism setting.

Animal agency in organizations

Despite the gradual transformation of leadership from single individuals to many individuals, this development has been almost unexceptionally anthropocentric. As a result, animals, in line with other non-human actors, continue to be seen "as mere black boxes and as placeholders: entities that stand in for more real and meaningful actors," (i.e., humans – authors' comment) "entities that merely relay force, a force from elsewhere, and direct it in a certain direction" (Sayes, 2014, p. 137). At the same time, agency is the ability to make a difference in the course of other agents' actions, whether this actor is human or non-human (Latour, 2005, p. 71). This *ability to make a difference* can be easy to underestimate sometimes but is difficult to ignore. For example, Sage et al. demonstrated how [mere] frogs can challenge if not stop a large construction project through their "different, even disorganizing, certainly troublesome, but never monstrous or simply managed, animal agencies" (2016, p. 447).

Among other animals, dogs and horses have also been given attention in the literature as companion species and working animals (e.g., Cunha et al., 2019; Haraway, 2003, 2008; Keaveney, 2008; Spiggle, 2008). Organizational literature has examples of the cat(s) *steering* an airport (O'Doherty, 2016) and the dog(s) *governing* a nation (Scoglund & Redmalm, 2016). The shift towards non-humans in organizational studies addresses animals not as different others but instead as becoming "increasingly center stage" participants that "act as a medium but also stimulus through which a whole range of activities and behaviors could be interpreted and understood" (O'Doherty, 2016, p. 419). As a result, they work as "an element in an assemblage of heterogenous parts out of which action is emergent" (O'Doherty, 2016, p. 425) and are distributed among different actors.

Animals in tourism

Tourism is based on the pursuit of personal satisfaction, and the desire to see and interact with animals can be a motivating factor for travel. Animals contribute to tourism in multiple ways: as attractions in their own right, as forms of transportation, as symbolic destination icons, and as components of regional cuisine (Markwell, 2015). Our understandings of animals are molded largely by cultural representations of them, and many of us pack these understandings with us when we travel (Markwell, 2015). At the destination, animals are transformed into tourism products for consumption, either metaphorically or materially, by tourists. Burns (2015) notes that objectifying animals as a resource available for consumption firmly places animals as the underprivileged counterparts in power relationships in tourism.

The perspective of animals as objects raises questions about the roles of animals in tourism experiences and about what they are doing during the time and space that frames the commercial experience: Are they seen as resources in the production of the tourism experience or as beings with intrinsic value? Are they seen as both? If they are considered a resource, would it be in the form of an operand (tangible) or operant (intangible) resource? The animal as a being with a body is often seen as an operand resource that lacks agency and provides goods (meat, other bodily products) or services (labor, aesthetics, entertainment) that are controlled by a human. However, animals do have agency and can contribute resources to the tourism experience, for example, their knowledge, skills, and personality. Assuming that animals have agency follows from an embodied ontology

or worldview that is at work in our understanding of animals. According to Steward (2009, p. 226), the following can be said about the concept of agency; an agent:
- can move the whole, or at least some parts, of something we are inclined to think of as its body
- is a center of some form of subjectivity
- is something to which at least some rudimentary types of intentional state (e.g., trying, wanting, perceiving) may be properly attributed
- decides when, how, and why its body is moving

Animals in tourism are selected for their cooperativeness. However, even then, their emotions, such as enthusiasm, stress, apathy, or happiness, influence tourists during the experience. Hence, animals cocreate the tourism experience. A relational ontology has been more frequently adopted in tourism research in recent years, based on the idea that humans are involved in the world as bodily beings through lived experience (Pons, 2003). In a discussion about the multi-relational model of tourism consumer experience, Lindberg et al. (2014) distinguished between the interactions tourists have with other human beings and with animals. How tourists relate to and interact with animals depends, according to the authors, on the behavior of the animal; on the tourists' skills, interests, and former experience; and on the cultural meaning, information and instruction provided by the guides.

In this chapter, we agree with Lindberg et al. (2014) that interactions between humans and other beings depend on several factors that acknowledge agency. For example, the behavior of an animal affects its relationship with humans, and the way the animal is looked at and understood by humans defines the relationship between the two beings. Guides often act as translators and facilitators of the relationship between animals and tourists. They have the most knowledge about the worlds of tourists and animals and can be seen as the bridge between these realms. Communication between humans and other animals is nonverbal, so the guide has the role of interpreting and translating this nonverbal language into words that tourists can understand. Seeing animals as beings with agency and resources in the tourism experience allows for the distribution of power. Additionally, when the animal is the actor with the most knowledge or skills, this will empower the animal and allow it to take a leadership role. Notably, among wild animals (where no human agency in place), leadership is very dynamic (Walraven, 2019), i.e., a leader either adapts to the circumstances or it will be naturally replaced.

Methodology

Despite the growing interest of organizational scholars on the role of animals in organizations (e.g., Connolly & Cullen, 2017; Cunha et al., 2019; Labatut et al., 2016), animal leadership in the tourism industry has only recently started to be addressed. One apparent reason may be some common methodological challenges associated with studying silent others such as animals. Hamilton and Taylor (2012, p. 44) put it as follows:

> Animals do not talk or write about what is going on in their lives nor do they wittingly contribute to the experience of work for humans [. . .]. This means that abattoirs, animal rescues helters, stables, veterinary surgeries and multiple other workplaces that have animals in-common add an extra layer of complication for researchers.

This constraint makes it especially important to examine the varied experiences of particular fieldwork(s). In our research, we used two cases – Russian huskies and Icelandic horses. Following others, we respected the crucial roles of these animals in tourist organizations, "we consider them to be 'actors' and certainly 'workers', whether they are aware of it or not" (Hamilton & Taylor, 2012, p. 46).

To study the phenomenon, we applied workplace ethnography, which has the power to reconnect researchers with the their study objects – in this case, Russian huskies and Icelandic horses – and facilitate "a deeper understanding of the world-views participants bring to our research studies as well as better recognition of our own personal assumptions that limit our research" (Zickar & Carter, 2010, p. 316). Since the focus of our study – the distribution of leadership between humans and animals – is an invisible issue, one may only see/trace the effects of it. By acknowledging the challenges related to communication between human and non-human actors, researchers should attach "subtle signifiers [. . .] to even the most mundane routines" (Hamilton & Taylor, 2012, p. 45). In this respect, the role of a researcher studying non-human agency is "to better attend to the minute displacements, translations, practices, riots, processes, protests, arguments, expeditions, struggles, and swap-meets – no matter what the actors involved may look like" (Sayes, 2014, p. 145), while keeping in mind that "even something as apparently mundane as washing the fur of a long neglected cat or dog becomes useful in the creation of a common work ethos" (Hamilton & Taylor, 2012, p. 45–46). Practically, this means that as researchers, we acted as tourists and followed the tourist experience in all possible details, e.g., leashing the sled dogs every morning and fastening them on the stake-outs for overnights, feeding the animals, and checking for injuries. Following daily routines as well as facing and discussing challenging situations while driving the sled dogs or riding a horse facilitates "apparently meaning-less duties and repetitions [. . .] to become highly meaning-full components of a carefully orchestrated collective performance" (Hamilton & Taylor, 2012, p. 43).

Data from the two cases were collected by the authors individually and analyzed jointly afterwards. One researcher took part in the horse riding in Iceland in two rounds, during 2014 and 2017, while the other one participated in the journey on the dog sledges in the Republic of Karelia (northwestern Russia) in 2017. Each data collection round lasted from several hours (horse riding) to several days (dog sledding) and allowed the researchers to experience variations in landscape, temperature, wind, surface, and other factors. In addition to the researchers themselves, the horse-riding tour included 10 people (among whom two were the tour guides) and 10 horses. The dog sledding experience was organized by 15 humans (two dog experts, a photographer, a doctor, several students, and representatives of Life Saving Services) and 32 dogs (mainly Siberian huskies). In addition to the researchers' own experiences, data were supported by informal interviews, observations, discussions, photo and video materials, as well as examination of relevant documents on the topic, such as those produced by the Mushing Federation of Russia and the travel notes of famous polar expeditions.

Due to the extreme climatic conditions of the research context, it is worth further considering its implications. Grint (2010) claims that "the sacred nature of leadership is not so much the elephant in the room but the room itself – the space that allows leadership to work" (p. 89). Animal-based tourism takes place in nature or even in the wild, offering a unique room for leadership. Taking into consideration the conditions necessary for distributed leadership, such as a dynamic environment (Thorpe et al., 2011), the question is how traditional leadership "survives" in conditions where any decrease/increase in the air temperature may revise the situation entirely. The answer is that it does *not* survive. The Arctic contexts where dog sledding and horse riding take place are constantly dynamic. To address this, we adopted, as Vásquez (2013) calls it, a "performative view of space," i.e., "one based on the interactions that involve heterogeneous actors – human and nonhuman," that can be imagined "as the coexistence of multiplicity, of heterogeneity" (p. 128), or a coexistence of trajectories as "stories-so-far." In practice, this means, for example, that a researcher should address the totality of dog sledding as a phenomenon:

> Dog sledding as a phenomenon is constituted not only through experience and knowledge in the relatively limited time span of the mushing itself but also through a juxtaposition of daily routines, logistics, obligations, worries and so on – elements that deeply permeate and shape the everyday life of the mushers. From this perspective, mushers' perceptions and experiences of dog sledding must be perceived as the result of a rather complex chain of actions and circumstances, shaped and embedded in and by individual life contexts.
>
> (Mosgaard et al., 2019, p. 287)

The narratives

Russian huskies

The history of dog sledding tourism in Russia is quite young and can be considered a further development from sled dog racing. Sled dog racing as an independent activity is very resource-intensive in terms of money and time, so it is often integrated with running dog sledding tours for support. The history of sled dog racing is associated with North America. In Russia, before the 1990s, these activities were considered only as part of national celebrations in northern regions as a tribute to local traditions. At the same time, the first gear for dog sledding were found on the Russian Novosibirsk islands and have been dated at approximately 8,000 years old. In some remote areas, dogs have always been a driving force for the development of northern territories and an irreplaceable part of everyday life, e.g., for communicating, hunting, and rescuing.

Although it is a type of recreational outdoor activity, dog sledding does not require special physical training and, in this respect, is available for all age categories of tourists, from children to pensioners, including people with reduced mobility. However, the role of the tourist in the experience (as a musher on the steering column or sitting along as a passenger) is based on the tour guide's judgment of the situation more than on the desire of the tourist. However, as this chapter is focused on distributed leadership in human-dog relationships, we will focus on the interactions between the musher (be it a tourist or a guide) and the dogs on the sled team.

A typical dog sledding trip with a passive tourist function is usually 15–25 km with several preselected stops for coffee breaks, which is a popular option for families with kids. For such trips, the route is usually pre-established (good track, fixed stops, no obstacles, etc.), and even an inexperienced person will sometimes be allowed to try the musher's role on a short segment of the route since the dogs know what to do. However, it is a different story when a trip is arranged over the course of several days with overnights in a cabin or in a tent. In these cases, the dogs given to a tourist become theirs for several days. Being a musher for several days gives the tourist a unique chance to learn and perform all relevant activities/routines/functions, including leashing, making stake-outs, feeding, driving, and so on and to learn to interact with the dogs in varying conditions in their natural habitat. Each of these practices contribute to the musher learning what kind of actors the dogs are, i.e., their strengths and weaknesses, as well as understanding their emotional profiles. All of this information is taken in by the guide and will be important when regulating the speed and amount of rest during the journey; it literally affects when to tighten the reins and when to leave them completely loose.

Sled dogs admire humans and would never hurt them – a fact that also adds to the attractiveness of the dog sledding activity for tourists. At the same time, while domesticated, they are not pets, as they demand the skill and experience of the tour guide in regard to organizing teams for work and rest. Each dog in a sledge is an important part of the team and is given its place based on its physical and social characteristics. While all dogs in the team are trained to pull the sledge, there is a lead dog that is usually more experienced in regard to selecting a route and/or organizing the work of others even when exhausted. Hence, although each dog has a name and can be directly addressed by the musher, the whole team is often steered via the alpha dogs, which set the pace for the team and discipline the free riders.

The musher's role in building their own experience is hard to overestimate. They are literally in the position of power to start and stop the movement of the whole team by means of the two commands, namely, "allez" (run) and "stop." However, sometimes, such as when going uphill, the musher needs to help the dogs by pushing the sledge. When crossing a frozen lake, the musher may need to make a detour from the planned route due to an ice crack or after talking to local fishermen about the safe areas of the lake surface. The tour guide (expedition leader) said the following:

> It is a myth that huskies choose the safest route. Individually, yes. However, they are not able to take into consideration the weight of the whole team including the sledges, cargo, musher, and other dogs.

Similarly, when moving through an ice field, the dogs are able to find a place to squeeze between the ice packs and to make it down the hill slowly. However, the musher has the responsibility of ensuring that the sledges and passenger can follow safely and will not get stuck or drop off the edge. The musher must be able to hold the bar and prevent the dogs from running away. This task is sometimes so challenging that the musher must simply allow the dogs to find the way. For example, in a forest, it may not be possible to use either the brakes or the legs to maneuver, such as when the snow is so deep and soft that the brakes do not help. While knee-deep in snow, the musher risks dropping the bar and losing the sledge. Similarly, the musher may just rely on the dogs' instincts and senses to detect wild animals (bears, wolves, lynx, moose) when camping overnight in the wild. The dogs protect the humans during the night, but the musher also has the duty to protect the dogs when moving along the highway or through the villages. Village dogs running free pose a risk for sled dogs, as they sometimes attack.

The examples above emphasize the role of gear in human-sled dog interactions. The technology of dog sledding – design/construction of the sledges, leashing methods, leashes, and so on – all matter in distributing leadership between the musher and the dogs. However, to an extent, the opportunity for distribution is provided and limited by such gear. The sledge is designed with regard to the landscape it will be used in, i.e., the lower body is more stable but is difficult to use in deep snow or on an uneven surface. Likewise, the way the dogs are leashed affects the use of the sledge. Hence, in human-sled dog interactions, leadership is always limited/conditional, which raises an interesting question: what is distributed, and how is it distributed? In the distribution of work, who is the leader? Is it the being who pulls, or the one who relaxes? Is it the person who can talk and yell? Is it whoever knows best? Or is it simply the being who leads? What contributes to the dynamics that might make the musher behave differently depending on the situation? Sometimes the musher must be a leader, navigating and controlling with words, hands, and actions; in other situations, the musher can give control over to the dogs and just enjoy the ride without intervention.

Icelandic horses

In Nordic, Arctic, and sub-Arctic tourism destinations, there are many types of commercial horse-based tourism experiences. Iceland, for example, is a country famous for riding trips with Icelandic horses, which are the only horse breed found on the island. These sturdy, small horses are descendants of the horses brought to the island by the Vikings in the Middle Ages. To protect the breed, no other horses are allowed to enter the country. For the last 30 years, international tourist numbers have grown significantly in Iceland, and the Icelandic horse has served both as a unique selling point and a marketing image for tourism promotion (Einarsson, 2010). A distinction is made between horse trekking, which is a trip that takes more than 24 hours, and horse rental, which refers to a horseback riding session that is at most a day trip. The relationship between horses and people is and has been strong and long-lasting, which is not surprising given that riding is literally an elevated state for a human being (Helgadóttir, 2006). However, the feeling of becoming elevated when climbing onto a horse is certainly not natural for most people, especially when they live in large cities, far away from large animals. Nevertheless, many tourists who come to Iceland want to try a riding trip.

A typical horse-rental trip lasts between one and four hours. Most of these trips can be found around Reykjavik. The stable often offers a pick-up service at the hotel, and groups are large, with up to 30 horses and riders. The tourists arrive at the stable, where they are briefed by the guides about safety and the practicalities of the trip. They are offered warm overalls and helmets for the chilly Icelandic climate. The horses that are coming on the trip have been saddled and are waiting in the stable. The guides ask how much riding experience their guests have and select a horse for every rider, often based on physical characteristics of both rider and horse (size, weight, etc.), considering the temperament of both humans and horses as important factors. The guide makes the decision based on the first impression of the guest and their knowledge about the characteristics of the horses. For example, nervous people get paired with quiet horses. People can get nervous when they come into the stable because it is overwhelming to be confronted with so many horses. Even though they are small for horses, Icelandic horses are still large animals. The stable manager of a large horse-riding company around Reykjavik shared the following story:

> We had an old lady coming here twice, but I did not remember her; we are quite busy. I gave her the same horse the second time she was here, this truly sweet horse that I knew was going to take good care of her.

After the riders and horses are matched, they gather in the paddock, where the tourists are assisted in mounting the horses. Most people joining these trips have no experience with horses or riding and are anxious about falling off and hurting themselves during the trip. The role of the guide is to explain to their guests as well as possible what to expect, how to act, and how the horses will react. In the interviews, the tourists reported that they rely on the experience of the horses; the horses know what to do because they have been doing it every day for years. The horses will follow each other, so the guide has to make sure he/she is in front leading the herd. Another guide will ride last to follow what is happening in the group. On these beginner trips, the horses will not speed up; rather, the aim is to go for a walk in the Icelandic landscape. The horses know the terrain and do not need the rider to guide them. They can also tell if the rider is experienced, and they tend to follow and react to more experienced riders while ignoring the messages of inexperienced riders. Hence, the direction and speed are decided by the lead horses, while the others act as followers. Short tours are offered all year round, but the guides have to assess if the weather is adequate for the trip to proceed. The stable manager explained the following:

> If it storms, we cancel the tours because the horses get nervous when it is truly windy, and this can be dangerous. Additionally, when the snow is truly wet, we have to cancel because it piles up on the horse. However, we go out in rain and in cold winters; we have protective gear.

Decisions to cancel the tour are based on the guides' judgment on whether they can trust the horses or not in certain weather conditions. In stormy weather, the animals are more unpredictable. Despite precautions and training, accidents sometimes occur. Horses are herd animals, and if one gets scared and starts running, the others will follow. It is the role of the guide to prevent the whole herd from running off by keeping the distances between horses short so they feel close to their herd. Nevertheless, the stable manager told the following anecdote about a recent accident during one of the trips:

> There was this girl that fell [off] of her horse and broke her hand. She had to go to the hospital, and I think she even missed her flight home. I went to visit her, and I felt terrible, but the only thing she asked me was if the horse was okay, because she fell off when the horse tripped.

The tourist in this case cared about the horse, and while she was unlucky to fall off, she did not blame the horse or guide for her situation. She knew it was an accident and that there are risks involved in these activities.

Furthermore, horses need care and protection when working in the tourism industry, as they become bored mentally from doing the same walk every day. They need time off from their tourism job, and they communicate this to the humans in the company in the following way:

> If you are let them out of the stable and they are not excited to go out. They start playing with the fence or just stand there. They stop responding to the environment, and then you know you need to do something. Of course, we always try to give them vacation before it comes to that. You see it in everything they do, similar to how you would see it in a person. If you see [. . .] they don't necessarily ask me to tell you that they are sad, but you see it.

The horses communicate with humans, and it is the role of the guide to translate the horse language for tourists. For example, horses can communicate that they are not happy with the way they are being treated. An Icelandic entrepreneur who has a farm and offers horse treks in summer explained the following:

> Sometimes the guests get angry about something, and this happens mostly with riders who think they can ride. This lady said, "This horse is just banging her head!" I said to her, "It's no wonder. I would also bang my head if you sat like that." It is difficult to say something because they are your customers, but you still have to take care of your horses, and you have to figure out how to tell the people they should handle the horse differently.

Sometimes the animals are rebellious and uncooperative when they do not want to walk or work for tourists. A guide in the north of Iceland told the following story of a riding tour that was very tiring for her because she had to take one of the horses as a hand horse (a horse led by hand) because it did not want to move forward with its rider:

> The horse just wouldn't move forwards because the rider was not doing anything. That horse moves forward when she feels the rider wants her to move forward. However, the lady said that she did not want to tell her what to do because she loves horses and she did not want to mistreat them.

During longer tours (more than 2 hours), the group of people and horses needs a break. The guide finds a good spot for a drink and a snack while people get off their horses to give them some rest. During these breaks, the humans are connected to the horses in a different way – they can look each other in the eyes and have a different type of contact. It is a moment of bonding, of getting to know each other, and the guide commented that this often helps people to feel more relaxed, as the break fosters the feeling that they can trust their horse.

Concluding discussion – Distributed leadership during animal-based tourism experiences

The purpose of this chapter was to explore leadership in human-animal interactions in tourism contexts. Animal-based tourism that takes place in natural surroundings may be referred to as a constantly dynamic environment and therefore should be welcoming of distributed leadership. The narratives presented in this chapter confirm that there is a place for distributed leadership in these animal-based tourism experiences. The question is not whether to consciously distribute leadership or not (Benson & Blackman, 2011) but rather whether the shifting and re-distribution of leadership throughout the duration of an experience is a natural and evolving phenomenon resulting from interspecies interaction and collective practice. Specifically, in the cases of Russian huskies and Icelandic horses, the issue of harsh and unpredictable Arctic conditions is also important for understanding the role of distributed leadership. The Arctic magnifies the need for an active role of animals in decision-making and leadership. In particular, the powerlessness of guides and tourists may function as a prerequisite for distributing leadership and trust to animals (for example, when animals instinctively know the terrain or potential hazards better than humans). Similarly, powerlessness may also be an indicator of a power shift, which acknowledges that humans willingly grant the lead to the animals. This requires trust and experience in the relationships between animals and humans in the context of tourism.

Human-animal interaction is mediated and structured through formal organizations; in our cases, the tourists went on tours in the Arctic wilderness together with animals. Our narratives do not capture the perspective of the animals involved due to a problematic language barrier (Willis, 2013). However, we understand them as actors and as workers, and as such, animals as collaborative leaders are constructed through human activities and processes that create meaning. We relied on our human participants during our research when they claimed understanding of the animals through interpretation of their behaviors, intentions, and personalities. We asked ourselves how leadership is distributed between humans and animals in the tourism experience. Our research question can essentially be boiled down to a quest to understand the human relationship with other non-human actors in the context and practice of tourism experiences. Haraway (2008) pointed to the problematic construction of dualism, or the distinction between man and animal, when understanding what happens when species meet. We also tried to deviate from anthropocentric assumptions when understanding the role of animals in organizational life. To answer the research question, we have organized the discussion into sections addressing the distribution of leadership between humans and non-human actors, the changing roles of leadership and followership when engaged in practice, and animals as leaders.

Distribution of leadership

An essential assumption made in this chapter when addressing the distribution of leadership between human and non-human actors is that we cannot take human domination over other animals for granted. The power relationships between humans and other animals are complex and to understand them, we have to release the human desire to control through knowledge. In fact, when humans know less about a particular situation, the distribution of leadership occurs in a process of releasing control and relying on trust. When engaged in the practice of the tourism experience, leadership was distributed in the tripartite guide-animal-tourist relationship. Hence, ontologically, we have placed animal actors on the same level as human actors, escaping the understanding of animals in organizations as only a source of mediation that enables social relationships to be created and stitched together (Hamilton &

Taylor, 2012). Instead, animals are part of the web of relationships. The animals in our stories have been able to subtly manage their human partners, thereby depending on the empathy of their human handlers, which is developed through understanding and addressing the needs of a non-vocal actor.

It is therefore important to note that power is not distributed evenly, as humans are ultimately responsible for the care of domestic animals. The power hierarchy is based on responsibility, and the tour guide is the initial leader that ultimately controls when to distribute leadership and to whom (humans, as well as non-humans). For example, the tour guide decides who is going to be the lead dog during a tour or which horse to give to a rider. In the timeframe of the tourism experience, this happens in the beginning when people arrive at the premises, obtain information about the trip, and prepare for departure. The animals have a more passive, awaiting role in which the tour guide prepares them for the trip with gear (saddles, sledges, etc.). However, the animals do have agency through their behavior and communication; they thereby rely on human empathy and on the ability of their handlers to understand animals. When horses are tired or bored of the trips, they convey this message through their behavior. It is up to the tour guide to decide whether to take the horse on a trip or not.

Changing roles

We observed that during a tour, leadership is dynamic and changes between participants. The guide has a leadership role based on their knowledge of the environment, the animals, and the design of the trip. The tourist, as the human handler of an animal, is often the group member with the least knowledge and experience. This means that the animal takes the lead on where to go and what to do. However, the human handler practices leadership through gear. The dogs are harnessed and connected to the sledge, which limits their possibilities for action. A horse has a saddle and bridle that is controlled by the human riding them. Although animals have more knowledge and better senses for decision-making when in nature, they do not have an equal role in controlling the situation. The guide and human handlers therefore have to pay attention to animal communication and transfer leadership when needed. Identifying the situations and moments when leadership has to be distributed to the animals requires experience and knowledge and should therefore be the role of the tour guide during the tourism experience. In the end phase of the experience, leadership is often distributed from tour guides to tourists. During the experience, the tourists have gained confidence and knowledge about the animals and are able to, for example, lead them back to the pasture or take off their harnesses.

Hence, the distribution of leadership is a practice that changes over time, where several leader-shifting moments can be identified. The communication of the intentions of the leader and the shift from one leader to the next depend on verbal and, most importantly, nonverbal communication. For example, in difficult terrain, dogs and horses have more responsibility and leadership in choosing a path. The musher and the rider rely on the instincts and senses of animals that are better developed than human senses in such settings. During an overnight stay in the wilderness of northern Russia, dogs protect humans from predators approaching the camp. In the same way, dogs may rely on protection from humans when crossing Russian villages, where dogs often roam free and pick fights with sled dogs. The musher has to take the lead by choosing the safest routes, and the passenger has to make sure that village dogs cannot approach. Therefore, leadership and responsibility are about caring for the well-being of others. The humans or animals being cared for trust the leader to care for their well-being. Mutual respect and trust are essential social capital during animal-based tourist trips because both animals and humans have to rely on each other for their safety and well-being. The amount of trust that a guide has in his or her animals depends on the situation. For example, horses who are not willing to work or are hot-tempered cannot be trusted the same way as happy, collaborative, and calm animals. In this case, the guide as a leader has to distribute leadership to the tourist that is riding this horse, and as such, takes a risk because they do not know the amount of prior experience and knowledge a visitor has regarding horses.

Animals as leaders

Connolly and Cullen (2017) offered an investigation of animal-related articles to support and argue for enhancing the visibility of animals within organizations. Acknowledging that animals have active leadership roles during the practices of the tourism experience is in line with the ambition of making animals more visible in the management and business literature. Through our ethnographic research, we observed that trust in animals and the ability to cede control are essential in the distribution of leadership between animals, guides, and tourists. Only in a leadership role can animals use their knowledge and experience for the common good of the group. However, for the well-being of animals and tourists, it is also essential that the guide is aware of shifts in leadership and is able to regain control when necessary. Communication with the animal is often a tacit skill of the guide that is developed through working with animals on a daily basis. The better the guide knows the individual animals, the better the experience will be for the tourists. Hence, to provide a valuable animal-based tourism experience, there must be a considerable amount of social capital in the form of trust, knowledge, and experience.

Specifically, Connolly and Cullen (2017) discussed how to consider our relationships with animals through the ethics of care. This ethics is based on direct and concrete relationships rather than on a set of principles designed to guide behavior. Care is considered both a value and a practice, and the ongoing interdependent relationships are emphasized as sites of care (Connolly & Cullen, 2017). In light of this chapter, we can understand the tourism experience as a site of care in which interdependent relationships between humans and other animals play out. During the tourism experience, concrete caring-for is focused on the welfare, protection, or enhancement of the cared-for; intrinsic value is placed on the life of the cared-for, rather than seeing them as the means to an end. Here, both tourists and animals are at risk of being viewed not from an intrinsic perspective but from an instrumental perspective, in which care is offered based on the useful services the cared-for may provide. Tourists pay for the trip, and therefore, the guide protects them and offers them an experience. This way of thinking most often occurs on short trips, such as the horse-riding tours presented in our narrative.

In particular, our narratives demonstrate that the intrinsic value of others is developed over time in relationships where the participants are developing interdependencies and getting to know each other. In this respect, the dog sledding example offers a possibility for understanding how concrete, intrinsic care is developed in a multispecies group. We contribute to the leadership and care discussion by arguing that animals are not only on the receiving end of care. Instead, when in a leadership role, given responsibility and power, animals care for their human companions. We let animals care for us when we acknowledge our own cognitive or sensory shortcomings. Animals are empowered and play a role in organizational life; their role as leaders and caregivers involves the ability to make decisions and judgments that contribute to the well-being of the multispecies group, which has not been acknowledged thus far. "People enjoy the satisfaction of communicating with another species and prize the feeling of connection to wildlife and nature" (Keaveney, 2008, p. 447) – in this respect, animal-based tourism is expected to benefit from distributed leadership precisely due to this bond.

References

Benson, A. M., & Blackman, D. (2011). To distribute leadership or not? A lesson from the islands. Tourism Management, 32(5), 1141–1149.

Burns, G. L. (2015). Animals as tourism objects: Ethically refocusing relationships between tourists and wildlife. Animals and tourism: Understanding diverse relationships, 67, 44.

Burns, J. (1978). *Leadership*. Michigan: Harper and Row.

Connolly, L. & Cullen, J. (2017). Animals and Organisations: An Ethic of Care Framework. *Organization and Environment* 31(4), 406–424.

Cullen-Lester, K. & Yammarino, F. (2014). Special issue on collective and network approaches to leadership. *Leadership Quarterly* 25, 180–181.

Einarsson, S. Ö. (2010). The role of the Icelandic horse in Icelandic history and it's image in the Icelandic media (Doctoral dissertation).

Cunha, M. P. E, Rego, A., & Munro, I. (2019). Dogs in organizations. *Human Relations*, 72(4), 778–800.

Fennel, D. A. "What's in a name? Conceptualizing natural resource-based tourism." Critical Debates in Tourism. United Kingdom: Channel View Publications (2012): 300–304.

Fitzsimons, D., James, K.T. and Denyer, D. (2011). Alternative Approaches for Studying Shared and Distributed Leadership. *International Journal of Management Reviews* 13, 313–328.

Gibson, J. J. 1979. The Ecological Approach to Visual Perception. Houghton Mifflin, Boston, MA.

Grint, K. (2010). The Sacred in Leadership: Separation, Sacrifice and Silence. Organization Studies, 31(1), 89–107.

Gronn, P. (2003). *The New Work of Educational Leaders: Changing Leadership Practice in an Era of School Reform.* SAGE Publications.

Hamilton, L., & Taylor, N. (2012). Ethnography in evolution: adapting to the animal "other" in organizations. *Journal of Organizational Ethnography*, 1(1), 43–51.

Haraway, D. J. (2003). *The Companion Species Manifesto: Dogs, People, and Significant Otherness.* (Vol. 8, Prickly Paradigm). Chicago: Prickly Paradigm Press.

Haraway, D. (2008). *When Species Meet.* (Vol. 3, Posthumanities). Minneapolis, Minn: University of Minnesota Press.

Harris, A. (2009). *Distributed Leadership: Different Perspectives.* Springer Netherlands.

Harris, A., & Spillane, J. (2008). Distributed leadership through the looking glass. *Management in Education*, 22(1), 31–34.

Helgadóttir, G. (2006). The Culture of Horsemanship and Horse-Based Tourism in Iceland. *Current Issues in Tourism*, 9(6), 535–548

Keaveney, S. M. (2008). Equines and their human companions. *Journal of Business Research*, 61(5), 444–454.

King, A. J., Johnson, D. D., & Van Vugt, M. (2009). The origins and evolution of leadership. *Current biology*, 19(19).

Kirk, P., & Shutte, A. M. (2004). Community leadership development. *Community Development Journal*, 39(3), 234–251.

Labatut J., Munro I. & Desmond J. (2016). Animals and organizations. Organization (London, England) 23(3): 315–329.

Latour, B. (2005). Reassembling the Social: an Introduction to Actor–Network–Theory. Oxford: Oxford University Press.

Lindberg, F., Hansen, A. H., & Eide, D. (2014). A Multirelational Approach for Understanding Consumer Experiences Within Tourism. *Journal of Hospitality Marketing & Management*, 23(5), 487–512.

Markwell, K. (2015). *Animals and Tourism: Understanding Diverse Relationships.* Channel View Publications.

McGehee, N. G., Knollenberg, W., & Komorowski, A. (2015). The central role of leadership in rural tourism development: a theoretical framework and case studies. *Journal of Sustainable Tourism*, 23(8–9), 1277–1297.

Meijer, E. (2019). Animal Languages: The secret conversations of the living world: John Murray Press.

Milne, Alan Alexander (1926). *Winnie-the-Pooh.* London: Methuen & Co.

Mosgaard, S., Butz, C., & Rune Waaler, R. (2019). Dog sledding as a field and research. Shedding the light on the research gaps. In Waaler, R. & K. Skjesol. *Dog Sledding in Norway: Multidisciplinary Research Perspectives.* Zürich: LIT Verlag, (pp. 277–301).

O'Doherty, D. P. (2016). Feline politics in organization: The nine lives of Olly the cat. *Organization*, 23(3), 407–433.

Pons, P. O. (2003). Being-on-Holiday: Tourist Dwelling, Bodies and Place. *Tourist Studies*, 3(1), 47–66.

Sage, D., Justesen, L., Dainty, A., Tryggestad, K., & Mouritsen, J. (2016). Organizing space and time through relational human–animal boundary work: Exclusion, invitation and disturbance. Organization, 23(3), 434–450.
Sayes, E. (2014). Actor–Network Theory and methodology: Just what does it mean to say that nonhumans have agency? *Social Studies of Science*, 44(1), 134–149.
Senge, P. M. (1990). The art and practice of the learning organization.
Senge, P. M. (2010). *The Fifth Discipline: The Art & Practice of The Learning Organization*. Crown.
Skoglund, A. & Redmalm, D. (2016). 'Doggy-biopolitics': Governing via the First Dog. *Organization*, 24(2), 240–266.
Spiggle, S. (2008). Making connections: Partner quality, identity, and community. *Journal of Business Research*, 61, 455–456.
Spillane, J. P. (2012). *Distributed Leadership*. Wiley.
Spillane, J. P., Halverson, R., & Diamond, J. B. (2001). Investigating School Leadership Practice: A Distributed Perspective. *Educational Researcher*, 30(3), 23–28.
Steward, H. (2009). Animal Agency. *Inquiry*, 52(3), 217–231.
Thorpe, R., Gold, J., & Lawler, J. (2011). Locating Distributed Leadership. *International Journal of Management Reviews*, 13(3), 239–250.
Valente, F., Dredge, D., & Lohmann, G. (2014). Leadership capacity in two Brazilian regional tourism organisations. *Tourism Review*, 69.
Vásquez, C. (2013) Spacing Organization, or How to Be Here and There at the Same Time. In Organization and Organizing: Materiality, Agency and Discourse, edited by Daniel Robichaud, and Francois Cooren, Taylor and Francis. 127–149.
Walraven, E. (2019). *Wild Leadership: What Wild Animals Teach Us about Leadership*. New Holland Publishers Pty, Limited.
Willis, P. (2013). *The Ethnographic Imagination*. Wiley.
Woods, P. A. (2004). Democratic leadership: drawing distinctions with distributed leadership. *International Journal of Leadership in Education*, 7(1), 3–26.
Zaccaro, S. J. (2007). Trait-based perspectives of leadership. American psychologist, 62(1), 6.
Zach, F. (2016). Collaboration for innovation in tourism organizations: leadership support, innovation formality, and communication. *Journal of Hospitality & Tourism Research*, 40(3), 271–290.
Zickar, M. J. & Carter, N. T. (2010). Reconnecting With the Spirit of Workplace Ethnography: A Historical Review. *Organizational Research Methods*, 13(2), 304–319.

Marcus Hansen, Jillian M. Rickly, Nigel Halpern,
Scott McCabe

9 A working holiday: From home to destination with a guide dog

Introduction

Tourism research is increasingly concerned with the role of animals in the industry (see Carr & Young, 2018; Fennell, 2012; Kline, 2018a, 2018b; Markwell, 2015). Such research often engages ethical debates regarding the human-animal interactions facilitated by tourism, particularly as tourists encounter animals as part of the destination experience or specific attractions. This includes tours to view wildlife in their natural habitat (Carr & Young, 2018; Fennell, 2012), animals performing for tourists' amusement (Fennell, 2012; Markwell, 2015), pack animals that act as transportation within a destination (Bertella, 2014; Cousquer & Allison, 2012), and even as food in gastronomic experiences of places (Kline, 2018a; 2018b). In addition, tourism research on human-animal interactions is beginning to investigate the relationships that transcend the boundaries of home and away. Carr and Cohen (2009) observe the benefits to tourists' well-being when accompanied by a pet, and Carr (2017) contends that tourists increasingly expect their pets to be catered to as a member of the family.

Despite the growing interest in some areas of human-animal interactions in tourism, research has yet to take notice of the role of service animals (Rickly, 2018). This is important given that the industry itself is currently ill-equipped to serve them. In fact, the European Commission (2015) finds that those traveling with a service animal are among the least catered for segments in Europe's accessible tourism market. This is despite the simultaneous expansion of reasonable accommodation legislation around the world and the increasing utilization of service animals to cope with mobility challenges.

Service animals are broadly defined as animals trained to support humans, which includes providing assistance and support as well as performing specific tasks (Huss, 2009). While service animals are most often dogs, they can be other animals as well, including monkeys, miniature horses, and pigs, among others. Of particular interest to this chapter are assistance dogs, which are trained to mitigate a variety of impairments, including vision impairment, hearing impairment, physical disability,

Acknowledgment: The authors would like to thank Guide Dogs for their collaboration in this research and for the opportunity to work more closely with guide dog owners (and their guide dogs). It has been a fantastic learning experience. This research was funded by the Economic & Social Research Council (ESRC) Business Boost Funds and Impact Accelerator Funds. As per contractual agreement with Guide Dogs, this chapter was approved for publication on the 1st May 2020. Nevertheless, any errors that remain are the responsibility of the authors.

https://doi.org/10.1515/9783110664058-009

medical disorder response, as well as provide emotional and psychological support. Specifically, the focus of this chapter is guide dogs, the work they perform for and with their owners who are vision impaired, and how this human-dog partnership is utilized outside their everyday environment, particularly in a tourism context.

It has been well-established that guide dogs improve their owner's mobility within their local communities while also contributing to higher quality of life through greater independence, confidence, and social interactions (Audrestch, Whelan, Grice, Asher, England, & Freeman, 2015; Gravrok, Howell, Bendrups, & Bennett, 2018; Lloyd, La Grow, Stafford, & Budge, 2008; Tomkins, Thomson, & McGreevy, 2011). Moreover, research indicates a preference for guide dogs as a mobility aid, as opposed to a long cane or other alternatives (Glenk, Přibylová, Stetina, Demirel, & Weissenbacher, 2019). Importantly, guide dogs are most effective when working in familiar environments wherein they are familiar with specific routes and can better respond to potential hazards. However, while on holiday they encounter unfamiliar environments, limiting the work they can do. Nevertheless, guide dog owners sometimes prefer to be accompanied by their guide dog, even if their capacity to guide is limited, as guide dogs are more than mobility aids, but one part of an inter-dependence that develops with and through the capabilities and communications of both human and dog as partners (Haraway, 2003, 2013; Higgin, 2012).

Because service animals in the tourism industry, and assistance dogs specifically, have been thoroughly underexamined in tourism research, the findings presented here are part of a larger project that utilizes an exploratory approach as the first stage towards investigating the human-assistance dog relationship as it affects travel behavior and tourism experiences. Thus, in an effort to shed light on the experiential dimensions of traveling with a guide dog, this chapter draws upon interviews with guide dog owners to reconstruct travel experiences from home to the holiday destination, highlighting the work performed by guide dogs, the challenges of traveling with a guide dog, as well as the limitations of their abilities when in unfamiliar environments. As one of the first pieces of research to focus solely on the role of guide dogs in tourism, this chapter further uncovers key points in the journey that present common challenges to traveling with a guide dog, while expanding the perspective of human-animal interactions research in the field. We also hope that this project will be a call for research focused on human-animal relations and social justice issues to turn attention to assistance dogs in the travel and tourism sector.

Accessible tourism and traveling with vision impairment

Holidays are recognized as a critical feature of modern life, an avenue for individuals to improve their quality of life (Dolnicar et al., 2012) and bring positive emotions and life satisfaction (Evcil, 2018). The United Nations General Assembly (1948) recognizes the right to leisure and international travel as human rights under the Universal Declaration of Human Rights (McCabe & Diekmann, 2015). There have been debates about the extent to which these rights encompass tourism (Breakley & Breakley, 2013), and it is without doubt that some countries have interpreted them as social rights (McCabe & Diekmann, 2015).

However, the rights of people with disabilities are much more clearly defined. In 2006, the United Nations' Convention on the Rights of People with Disabilities (CRPWD) recognized that people with disabilities have a right to access services from all areas of citizenship, including transport and the built environment (Article 9), as well as all areas of cultural life (Article 30) (Michopoulou et al., 2015). More recently, the UN has included "Reduced Inequalities" as one of the 17 Sustainable Development Goals, expanding the importance of accessibility. Nevertheless, access issues for people with disabilities are well-documented within the tourism industry (Buhalis & Darcy, 2011; Mesquita & Corneira, 2016; Nyanjom et al., 2018), which has inspired the rise of accessible tourism (Buhalis & Darcy, 2011; Cohen et al., 2014; Vila et al., 2015). According to Buhalis & Darcy (2011, p. 10), accessible tourism is defined as:

> a form of tourism [. . .] that enables people with access requirements, including mobility, vision, hearing and cognitive dimensions of access, to function independently and with equity and dignity through the delivery of universally designed tourism products, services and environments.

As accessible tourism strives to include all people in tourism activities, the idea of reasonable accommodation becomes fundamental to the concept. According to the UN CRPWD (2006), reasonable accommodation is the necessary and appropriate modification of a service, which does not impose undue burden on the service provider, in order to ensure equal opportunity for enjoyment. Notably, this principle is captured in some disability legislation, such as the Equality Act 2010 (UK) and the Americans with Disability Act 1990 (USA) (Nyanjom et al., 2018; Veitch & Shaw, 2011), which also builds upon the UN's CRPWD in that they focus particularly on access needs stemming from the following dimensions of disability: mobility, vision, hearing, intellectual/cognitive, mental health, sensitivities, and other.

Further, the aforementioned legislation also recognizes the difference between disability and impairment. An impairment is the loss of physiological or psychological function, while disability is a lack of ability resulting from impairment (World Health Organization, 2019). In other words, disability is relational to the environment, as a person with an impairment need not acquire a "disability" if enabling environments are designed (Swain et al., 2013). Unfortunately, the diversity of impairment and disability, and their resulting access needs, are not always understood in the industry. Indeed, this lack of understanding is highlighted by Kong and Loi (2017) who find that destination management stakeholders often assume people with disabilities were not interested in traveling or they stereotype them as simply wheelchair users (McKercher & Darcy, 2018). Thus, by being wheelchair accessible, their assumption is that attractions are accessible to all disabilities, when in fact this demographic is incredibly heterogeneous. Indeed, Buhalis and Darcy (2011) argue that disabilities are classified into a number of different categories, with each requiring a different level of support which exists along a continuum based on the degree of ability of the individual.

Considering the growing interest in accessible tourism, broadly, it is perhaps surprising that research on traveling with a vision impairment has been largely neglected thus far (Devile & Kastenholz, 2018; Poria et al., 2011; Richards et al., 2010). The World Health Organization (2010) estimates that at least 314 million people live with low vision or are blind. In the UK, alone, over 2 million people are vision impaired, a figure that is estimated to increase due to an ageing population and other underlying health issues (Small, 2015). Small and Darcy (2010) find that people with vision impairment (PwVI) are much less likely to travel, as compared to other impairments, and this is even more so for international travel (Loi & Kong, 2017). Specifically, Poria et al. (2011, p. 152) identify the "fear of missing information," such as changes to flight details, train platforms, and so on, particularly in noisy environments, as a source of travel anxiety for vision impaired tourists. Understandably, independence and confidence have considerable bearings on travel behavior (Loi & Kong, 2017).

Despite the focus on the visual dimensions of tourism (Urry, 1990), it is an embodied experience (Palmer & Andrews, 2019; Small & Darcy, 2011). As a result, touristic experiences are individualistic and subjective. Thus, vision impairment does not diminish tourists' experiences, but rather results in a different experience in which tactile, auditory, and olfactory sensations become critical (Richards et al., 2010; Small et al., 2012; Small, 2015). Nevertheless, PwVI do have specific access needs. As tourism takes place in unfamiliar environments, guiding becomes key for PwVI, and wayfinding is of particular concern for independent travel (Small et al., 2012). As a result, they are more reliant on others for assistance, be they a carer, friend, relative, a guide dog, or even service staff (Small et al., 2012), and challenges related to wayfinding are likely to have a negative influence on their enthusiasm to travel (Bosch & Gharaveis, 2017).

While guiding PwVI is a form of assistance, it should not be assumed that requiring a guide diminishes the experience for vision impaired tourists. Richards (2013, p. 98) argues that "being guided is a physical and interpersonal act between two bodies which concerns primarily the senses of touch and hearing between the sighted guides, the vision impaired person and their surroundings." In a sense, the bodies experience the journey in relation to one another. Considering the embodied relationality of guiding, it becomes obvious that PwVI would develop a deep, emotional connection to their guide dog. Guide dogs do not arrive as fully trained mobility devices, but rather after undergoing their basic training they begin their person-specific training. The dogs learn to understand their owner's communication, and ways of communicating, while the person is simultaneously learning the dog's personality and responsiveness. As such, it is an evolving relationship that is continually unfolding with time and experiences (Higgin, 2012).

Unfortunately, despite growing reasonable accommodation legislation worldwide, research suggests greater access issues for PwVI when they are accompanied by a guide dog, particularly in transport, at visitor attractions, and in hospitality settings, such as taxis, museums and/or restaurants (Devile & Kastenholz, 2018; Lloyd et al, 2008; Matsunaka & Koda, 2008; Mesquita & Carneiro, 2016). Typically, these barriers are a result of a lack of awareness from the tourism industry in regards to the specific needs of PwVI and their guide dog (Devile & Kastenholz, 2018). Thus, it is not surprising that Richards et al. (2010) find that inclusive attitudes towards guide dogs led to safer and more enjoyable tourism.

Research design

Guide dogs for people with vision impairment

A guide dog is a domestic dog specially trained to aid PwVI with mobility by guiding them from A-to-B (Craigon et al., 2017). Guide dog training begins when the dog is 6 weeks old, although it does not formalize until it is 14 months, and then lasts an additional 20 to 28 weeks before the dog is paired with an owner (Price, 2017). Many of the working guide dogs are the result of specialized breeding programs, funded and managed by charitable organizations, such as The Guide Dogs for the Blind Association (UK) and Guide Dogs for the Blind (USA), among others.

Previous research has discussed the role of guide dogs from a day-to-day perspective (see Audrestch et al., 2015; Gravrok et al., 2018; Higgin, 2012), but has yet to focus on the role of a guide dog when traveling outside of its home environment. The role of a guide dog is to guide their handler safely past obstacles and hazards once given a direction. The dogs are trained to follow directional commands, stop at curbs, stop to avoid vehicles, and locate certain objects (Craigon et al., 2017). Thus, research has demonstrated a number of benefits to owning a guide dog in day-to-day life, such as increased mobility, independence, confidence, and social interaction (Audrestch et al., 2015; Gravrok et al., 2018; Tomkins et al., 2011). Indeed, more recent research indicates a preference for the use of guide dogs as a mobility aid, as opposed to the long cane or more modern alternatives (Glenk et al., 2019). Yet, challenges remain for guide dog owners as they struggle to locate appropriate facilities for their dogs and the role of the guide dog is also typically misunderstood by the general public (Mesquita & Corneira, 2016).

Importantly, the role of a guide dog is not simply that of a working dog. Only a fractional part of the time is spent "working," meaning the rest of the time the dog takes the role of a pet, thereby developing a uniquely close bond with its handler

(Craigon et al., 2017; Higgin, 2012). This presents real challenges and opportunities for the tourism industry, which is increasingly catering to pets while also obliged to offer reasonable accommodation to guide dog owners. Therefore, in an effort to shed light on the experiential dimensions of traveling with a guide dog, this chapter draws upon interviews with guide dog owners to reconstruct a travel journey from home to the holiday destination. The discussion highlights the work performed by guide dogs, as well as the limitations of their abilities and the barriers experienced in getting to and from the destination. As one of the first publications to focus solely on the role of guide dogs in a tourism context, this chapter uncovers key points in the journey that present common barriers while traveling with a guide dog, as well as demonstrating the human-guide dog partnership as much more than a mobility relationship.

Methods

This study sought to gather guide dog owner experiences of traveling with a guide dog, the role the dog plays in guiding its owner in unfamiliar environments, and the challenges they encountered in tourism settings. The focus is on overnight trips away from home in order to capture the tourism context specifically, rather than trips taken during everyday life. Qualitative approaches are considered useful and necessary in the context of research that focuses on the complexities of visitors' needs and experiences (Cetin & Biligihan, 2016; Ryan, 2000). Therefore, to gain a deeper understanding of the specific issues for this topic, semi-structured interviews were employed (Flick, 2014; Saunders et al., 2016). A total of 27 interviews were conducted, averaging 45 minutes in duration. Each took place over the phone and was audio-recorded for transcription and analysis purposes. All of the participants were current guide dog owners or had recently retired their dog and were on the waiting list for another. Over half of the participants were female, with ages ranging from 20s to 70s. Some had only recently acquired a guide dog (2 years prior), whereas others had guide dogs for 40+ years. To protect the participants' identity, anonymous identifiers were employed instead of their names, and similarly the guide dogs' names have been replaced with "guide dog."

Participants were sourced from the membership database of The Guide Dogs for the Blind Association (simply "Guide Dogs" henceforth), and interviews were conducted in conjunction with a survey as part of a larger study about traveling with a guide dog. Guide Dogs is a UK-based charity that provides guide dogs, as well as other services and technologies, to people in the UK who are blind or partially sighted. For the purposes of this chapter, findings are derived from responses to specific questions related to experiences with their guide dog: how they prepare for an overnight trip, how they got to their destinations, and how they experienced the destinations. The audio recordings were transcribed, and the resulting transcripts were analyzed through thematic analysis by two of the authors (Boyatzis, 1998), with the development of themes guided by a review of the disability, vision impairment, and accessible tourism literatures.

Findings

Preparing for an overnight trip away from home

While owning a guide dog increases levels of mobility, independence, confidence, and social interaction (Audrestch et al., 2015; Gravrok et al., 2018; Tomkins et al., 2011), much less is known about potential benefits, challenges, and limitations of traveling away from home with a guide dog. Due to this uncertainty, Guide Dogs has created an extensive travel checklist for those considering international travel with their guide dog (Guide Dogs, 2019). Arranged in the style of a "frequently asked questions," the checklist attempts to address all matters concerning international travel with a guide dog and particularly emphasizes the implications of travel for the dog's welfare. Importantly, it begins by questioning whether it is necessary to bring the guide dog along.

Based on our findings, traveling domestically with a guide dog is a common occurrence, whether for visiting friends or relatives, for holiday, or work purposes. However, the level of preparation required prior to travel depends on the type of trip planned. While travelling domestically requires extensive logistical planning, traveling abroad involves especially complex preparations. Veterinary records and the dog's pet passport are required. Certain vaccinations and treatments must be up-to-date, with some specific to the destination (UK Government, 2019). According to current UK regulations, guide dog owners are required to have their guide dogs vaccinated for rabies and have blood tests taken at least 30 days after the vaccination, then wait at least 3 months from the date the successful blood sample was taken before travel is permitted. Finally, the vet has to provide an animal health certificate (AHC) no more than 10 days before travel. Proof of microchipping, vaccination history and a successful rabies antibody blood test result must be carried when traveling (UK Government, 2019). It is important to the note that regulations do vary across countries, and this must be observed in addition to the UK regulations.

Participant 8, a guide dog owner of 7 years in their 50s, explained that for them, preparing for going abroad with their guide dog involves even more than vet visits and paperwork, but also communicating with service providers and confirming reservations:

> Letting the accommodation know, making sure that the airline were prepared for him [. . .]. And then getting all his paperwork done in time, his tapeworm treatment and the animal

> reception can be bad through Gatwick. And then in terms of his preparation, making sure that he had all of his home comforts, so his blanket and having to weigh his food out.

Further, because the role of the guide dog is more than that of a working dog but also that of pet, developing an intimate partnership with its owner (Craigon et al., 2017; Higgin, 2012), packing for a guide dog is akin to packing for any companion dog. Food has to be measured for dispensing on a daily basis, bags for cleaning up after toileting. Toys and other home comforts, such as blankets, are included as well to ensure the guide dog is comfortable and to help alleviate any anxiety when traveling. Some guide dog owners have different leads and collars for their dog, one for work and one for play, which are included. Towels are also brought for bedding and any cleaning that might be required.

The appropriate services, such as transport and accommodation, also need notifying. While it is illegal to refuse service to someone with a guide dog, notifying hotels prior to arrival is recommended in order to avoid negative encounters. It seems common practice is to book a room online first, then call the hotel afterwards to inform them they would be bringing a guide dog. Guide dog owners believe such practices are necessary, even among major accommodation brands, to avoid discrimination, such as being refused a room or other uncomfortable situations upon arrival. However, some participants highlighted how Hilton Hotels, in particular, offers a service dog option when booking online, although this seems the only accommodation provider known to do so.

Participant 26, a guide dog owner of 10 years in their 40s, described the planning behind their domestic overnight trip:

> Preparing [guide dog] for the trip, obviously there's the basics of packing his food, making sure he's got what he needs, food, blankets and so on. [. . .] What I did obviously I booked in advance with the train company, made sure I booked an extra seat so that [guide dog] had enough space and you know, we got assistance to Kings Cross. [. . .] Got assistance off the train in Edinburgh.

Thus, much preparation goes into planning an overnight trip with a guide dog, which is only made more challenging when traveling abroad. As a result, in addition to their other recommendations for travel, Guide Dogs specifically recommends notifying the airline at least 48 hours prior to travel. However, they also ask owners to strongly consider whether it is appropriate to undertake such a journey with their guide dog. Appropriateness of the guide dog for the destination situation is particularly important to the dog's welfare. This includes the distance and therefore the length of the flight, the weather – especially worrisome are warmer destinations than the dog might be accustomed to – but also potential cultural differences that might not view dogs as positively as the UK, and the extent of the legal system of the destination to support reasonable accommodation legislation must all be considered. If PwVI are accompanied by another person who can assist with guiding, it recommended that the person deliberate on whether the dog's limited guiding abilities in the unfamiliar environment are worth the potential stress and anxiety that travel may cause for the dog.

Getting to the destination

The United Nations' CRPWD (2006), as well as the UK's Equality Act (2010) recognize that people with disabilities have a right to reasonable accommodation in accessing services from all areas of citizenship, including transport, the built environment, and tourism (Michopoulou et al., 2015). As an example, under part 12 of the Equality Act (2010) it is specifically against the law for taxis and minicabs to refuse service to a guide dog owner traveling with their guide dog, unless the driver has a medical certificate for exemption. Yet, traveling by taxi can be notoriously challenging for guide dog owners, with many experiencing regular refusals. For example, Participant 16, a guide dog owner of 25 years in their 40s made the following comment:

> Taxis, I don't use black cabs unless I absolutely have to because it's just a lottery you know, turning up at the [taxi] rank knowing whether they're going to pick me up or not.

The issue relating to taxis is typically due to a lack of awareness and poor training for staff that is in fact common across the service sector when confronted with PwVI. Little accountability exists from local councils, meaning such discrimination is far from uncommon. As a result, Devile and Kastenholz (2018) called for the promotion of disability awareness programs to improve tourism policy and increase awareness. More broadly, negative experiences, such as taxi refusals, can deter travel for guide dog owners and may have further ramifications for longer travel journeys where timely connections to other transport modes need to be taken. Therefore, a lack of training and awareness of front-line staff creates critical barriers to travel with a guide dog.

Despite airlines having official policies on service animals (see for example, EasyJet's policies https://www.easyjet.com/en/terms-and-conditions/passengers-with-specific-requirements), front-line staff are not always aware of these or how to handle such situations. Participant 18, a guide dog owner of 3 years in their 30s, described their experience at an airport in which airline staff were unprepared for handling a guide dog's booking and were further unaware of the kinds of paperwork that were required to manage the situation:

> the airport, the check-in lady [. . .] tried to charge me for the dog's bag. When I told her I didn't have to pay for it, she then said I just need to ring. So she rang up someone and had a whispered conversation, then turned to me and said can you show me the dog's documentation. And I said well what would you like to see? And then she goes I don't know. [. . .] the fact the security didn't know what the hell was going on, the fact that the person helping me kept touching the dog.

However, traveling with a guide dog need not be so complicated and full of obstacles if well prepared for what to expect. Indeed, as some participants explained, it can seem a rather straightforward affair, especially when traveling a familiar route. Participant 21, a guide dog owner of 9 years in their 60s, for example, described how their recent overnight trip went:

> When we go down to our friends in Norwich, then it's going over to the bus with all the stuff in a rucksack or a case and that adds slightly logistically to it all because you've got your own stuff, you've got the dog's stuff and then it's getting on and off the bus, finding the train. Usually getting assistance to ring through if you've got to change trains. I usually end up booking a seat for the dog as well, which they'll do. So there's my wife and myself and then an area for the dog to be able to sit in front of as well.

Moreover, travel can be made simpler for guide dog owners when accompanied by a person not reliant on the guide dog. Participant 19, a guide dog owner of 11 years in their 60s, travels with their wife and described a rather easy holiday journey by car:

> You just go and load the dog up in the car and what have you and off you go. And depending how far you're going, around about two hours you try and plan for a stop so the dog can have a busy [toilet break] if they need it or stretch their legs. We'd normally do that at a service station. And you learn about service stations by experience because some of them have fairly large grass areas, some of them have designated dog spending [toileting] areas. Some of them are absolutely abysmal. So you learn that by experience. So you have a checklist of service stations, if they've got good facilities you go and use them, if not you find the next closest one that does.

Getting to the destination can present the guide dog owner with considerable challenges, despite pre-planning the trip, with challenges more prevalent for those traveling alone. Guide dog owners have experienced problems with most transport providers, be they taxis, trains, or airplanes.

Relief areas for toileting

While some guide dogs are able to last up to 12 hours without needing to toilet, most recommendations suggest adult dogs receive a toilet break every 3–6 hours. As such, toileting the guide dog and finding relief areas can become an issue when traveling, especially over long distances, and many transport hubs, such as train stations and airports in the UK and abroad, are not consistently equipped with toileting areas.

With regards to rail stations, one exception is Birmingham New Street in the UK, which developed its dog relief area in collaboration with Guide Dogs. The availability of relief areas is on the rise in airports; however, such areas are sometimes located prior to security, making them inaccessible immediately before boarding and when changing planes. Further, among those areas located inside of security checkpoints, many airports still expect guide dogs to spend on a designated area of the tarmac. This raises a number of challenges for the guide dog: the designated areas are typically in the vicinity of jet engines and other tarmac traffic, making the guide dog too anxious. Further, not all guide dogs are trained to spend on concrete and therefore prefer grass. While newer installations that place relief areas within the terminal are more convenient, there are complaints that these are "too sanitized" so that the dog does not want to toilet. Participant 13, a guide dog owner of 2 years in their 50s, explained how long-haul flights were challenging for their guide dog:

> I think the toileting is one of the bigger issues because you've got to find it yourself [. . .] when you fly with a dog [. . .] the airports don't cater that well for spending for them prior to the flight [. . .] they don't have a spending area. [. . .] they like you to check-in two and a half hours in advance [. . .] I think that is a bit of an issue that they ought to really resolve.

Similarly, Participant 5, a guide dog owner of 14 years in their 50s, also argued for designated toileting areas at airports:

> I think it would be good to actually see sort of designated relief areas at airports [. . .] I know at Liverpool John Lennon Airport, it's either where the people go for a cigarette or [. . .] on the runway. They probably should have a designated area for the dogs really shouldn't they.

Thus, the need for such basic facilities to be present at all transport hubs is about supporting the guide dog's welfare and complying with reasonable accommodation legislation. Further, the lack of basic facilities can contribute to anxiety among guide dog owners, due to concern for their guide dog's welfare, while adding to the extra planning involved with their travel and may even act as a demotivator.

At the destination

The biggest challenge for PwVI, once at the destination, is wayfinding (Bosch & Gharaveis, 2017). Recent research suggests that guide dog owners prefer to work their guide dogs, even in unfamiliar places, as they are perceived to be faster and safer mobility aids, as opposed to the long cane or technological alternatives (Glenk et al., 2019). Additionally, the very presence of their guide dog adds comfort, companionship, confidence, and a sense of independence. While even in unfamiliar places some guide dogs can be relied upon to avoid obstacles on the pavement, stop at curbs, and avoid cars, PwVI will still be reliant on additional support for wayfinding (Small et al., 2012), such as locating the correct platform at train stations or specific street addresses. Notably, some adaptive technologies, such as mobile GPS applications, can assist in wayfinding and support their independence in unfamiliar environments, contributing to their confidence to guide themselves to locations while on foot, such that assistance is only required upon arrival.

It is important to understand, though, that the PwVI-guide dog relationship is a dynamic one in which the communication between the two results in an ever-evolving partnership (see Higgin, 2012). As a result, guide dogs can become familiar with new routes and adjust to new environments, even temporarily. So while the role of the guide dog is reduced while accompanying their owner on holiday, additional support from destination front-line staff at the start means guide dogs can add to independence within a destination. At the hotel, for example, initial assistance is needed in showing the guide dog owner to their room and back to the reception again, as well as to a relief area for the dog and back again, and other similarly necessary locations. In such circumstances, the guide dog owner will typically also count the steps in both directions to ensure accuracy along the route. Once the person and guide dog are familiar with the route, they can navigate it independently. Participant 1, a guide dog owner of 37 years in their 60s, illustrated this process during a recent trip:

> Once I got there to the hotel, the Premier Inn, they were fine, they gave me a downstairs disabled room you know, they were quite happy to give me anywhere but she said being a ground floor, that way you know, it'll make life a lot easier for [guide dog] because it'll have more room for her. My friend showed me where the nearest park was or where I could free-run her and relieve [toilet] her. Also, my friend showed me the route from the hotel to the conference center. Once she'd showed me that there and back and where I could take [guide dog] at lunchtime to relieve [toilet] her and whatever, things slotted in nicely for the whole week.

Once settled at a hotel, typically, the guide dog owner will also search the room to ensure nothing is present that the dog might eat and cause illness. Indeed, the dog's behavior in unfamiliar environments also has to be managed, particularly in regard to eating. It is important to remember that the guide dog is still a dog, albeit a highly trained one. While it has been trained to work for its owner, it is also reliant on that person for its welfare. Indeed, Participant 26, a guide dog owner of 10 years in their 40s, spoke of the challenges the behavior of others and their dog presents:

> One of the big challenges is actually litter. I mean it sounds really basic but the amount of rubbish that's dropped on public transport. And of course, he's a Labrador, so he eats it and you know, it actually has made him quite ill.

By initially receiving assistance, guide dog owners are able to learn new routes, meaning once unfamiliar environments can become familiar and thus enable guide dog owners to explore these areas more independently. This demonstrates the ways reasonable accommodation inspired legislation acts to foster the three values of accessible tourism: independence, equity, and dignity (Darcy, 2006). A relatively minimal effort by front-line staff can have tremendous implications for PwVI traveling with their guide dog. Yet, the debate is ongoing concerning the efficacy of taking a guide dog on trips to unfamiliar places, as the dog's abilities are limited and travel may add further stress and anxiety into the experience.

Conclusion

The role of guide dogs within the travel and tourism industry remains an under-examined area of research. While there is burgeoning research on the role of pets in holiday decision-making and travel behavior, little is known about traveling with a guide dog. As such, this chapter provides first insights into traveling with a guide dog in a tourism context: the role of a guide dog when traveling away from home overnight has been outlined, as have the subsequent challenges experienced by both guide dog and owner. In their home environment, the use of guide dogs is the preferred option as a mobility aid for PwVI. However, this chapter demonstrates that guide dogs also have a critical role to play during travel to the destination and subsequently guiding their owner around the destination. While the guide dog offers practical contributions, the companionship provided is also significant to the guide dog owner when away from home.

The preparation for travel with a guide dog varies depending on a number of elements and can become quite complex. While Guide Dogs provides guidance on international travel, they also encourage owners to explore whether it is indeed necessary to take the guide dog abroad, due to their diminished role and the implications of travel for the dog's welfare. While the primary role of the guide dog is mobility support, it also provides comfort, companionship, and supports confidence, leading to a more independent life for the owner. This makes the decision of traveling with the guide dog or leaving her/him at home while away on holiday a difficult one for many guide dog owners, which can be further exasperated by access challenges encountered along the journey. Reasonable accommodation legislation supports access for guide dogs across the travel and tourism industry. Yet, much remains to be done to improve the consistency of services and staff training in this area, and doing so has important implications for assistance dog welfare more broadly.

References

Audrestch, H.M., Whelan, C.T., Grice, D., Asher, L., England, G.C., & Freeman, S.L. (2015). Recognizing the value of assistance dogs in society. *Disability and health Journal*, 8(4): 469–474.

pictures of animals in tourism promotion. *Tourism Recreation Research*, 38(3), 281–294.

Bertella, G. (2013). Ethical content of pictures of animals in tourism promotion. *Tourism Recreation Research*, 38(3), 281–294.

Bosch, S.J. & Gharaveis, A. (2017). Flying solo: A review of the literature on wayfinding for older adults experiencing visual or cognitive decline. *Applied ergonomics*, 58(2):327–333.

Boyatzis, R. E. (1998). *Transforming Qualitative Information: Thematic Analysis and Code Development*. Thousand Oaks, CA: Sage Publications.

Breakley, N. & Breakley, H. (2013). Is there a right to tourism? Tourism Analysis, 18(1): 139–148.

Buhalis, D. & Darcy, S. (Eds.) (2011). *Accessible tourism: Concepts and issues.* London: Channel View Publications.
Carr, N. (2017). Recognising the position of the pet dog in tourism. *Annals of Tourism Research,* 62(1): 112–113.
Carr, N. & Cohen, S. (2009). Holidaying with the family pet: No dogs allowed!. *Tourism and Hospitality Research*, 9(4): 290–304.
Cetin, G. & Bilgihan, A. (2016). Components of cultural tourists' experiences in destinations, *Current Issues in Tourism*, 19(2): 137–154.
Cohen, S.A., Prayag, G., & Moital, M., (2014). Consumer behaviour in tourism: Concepts, influences and opportunities. *Current issues in Tourism*, 17(10):872–909.
Cousquer, G. & Allison, P. (2012). Ethical responsibilities towards expedition pack animals: The Mountain Guide's and expedition leader's ethical responsibilities towards pack animals on expedition. Annals of Tourism Research, 39(4): 1839–1858.
Craigon, P.J., Hobson-West, P., England, G.C., Whelan, C., Lethbridge, E., & Asher, L. (2017). "She's a dog at the end of the day": Guide dog owners' perspectives on the behaviour of their guide dog. *PLoS One.*
Darcy, S. (2006). *Setting a research agenda for accessible tourism.* Sustainable Tourism CRC.
Devile, E. & Kastenholz, E. (2018). Accessible tourism experiences: the voice of people with visual disabilities. *Journal of Policy Research in Tourism, Leisure and Events*, 10(3): 265–285.
Dolnicar, S., Yanamandram, V., & Cliff, K., (2012). The contribution of vacations to quality of life. *Annals of Tourism Research*, 39(1): 59–83.
Easy Jet (2019). Passengers with specific requirements [online]. Available at: https://www.easyjet.com/en/terms-and-conditions/passengers-with-specific-requirements Accessed: 19/11/19.
European Commission (2015). Mapping and performance check of the supply of accessible tourism services. Brussels, Belgium. p. 104.
Evcil, A.N., (2018). Barriers and preferences to leisure activities for wheelchair users in historic places. *Tourism Geographies*, 20(4): 698–715.
Fennell, D.A. (2012). Tourism and animal rights. *Tourism Recreation Research*, 37(2): 157–166.
Flick, U. (2014). *An Introduction to Qualitative Research*. London: Sage Publications Ltd.
Glenk, L.M., Přibylová, L., Stetina, B.U., Demirel, S., & Weissenbacher, K. (2019). Perceptions on Health Benefits of Guide Dog Ownership in an Austrian Population of Blind People with and without a Guide Dog. *Animals*, 9(7): 428.
Gravrok, J., Howell, T., Bendrups, D., & Bennett, P. (2018). Adapting the traditional guide dog model to enable vision impaired adolescents to thrive. *Journal of Veterinary Behavior*, 24(1): 19–26.
Guide Dogs for the Blind Association (2019). International Travel Checklist [online]. Available at: https://web.guidedogs.org.uk/-/media/Project/GuideDogs/GuideDogsDotOrg/Files/Resources/Guide-dog-owners/Assistance-dog-travel-guide/international-travel-checklist.pdf Accessed: 19/11/19.
Haraway, D.J. (2003). *The companion species manifesto: Dogs, people, and significant otherness* Chicago: Prickly Paradigm Press, pp. 3–17.
Haraway, D. J. (2013). *When species meet* (Vol. 3). U of Minnesota Press.
Higgin, M. (2012). Being guided by dogs. In L. Birke & J. Hockenhull (Eds.), *Crossing boundaries: Investigating human-animal relationships*. BRILL. (pp. 73–88)
Huss, R.J. (2009). Why context matters: Defining service animals under federal law. *Pepp. L. Rev.*, 37(1): 1163.
Loi, K.I. & Kong, W.H., (2017). Tourism for all: challenges and issues faced by people with vision impairment. *Tourism Planning & Development*, 14(2): 181–197.

Lloyd, J.K., La Grow, S., Stafford, K.J., & Budge, R.C. (2008). The guide dog as a mobility aid part 1: perceived effectiveness on travel performance. *International Journal of Orientation & Mobility*, 1(1): 17–33.

Kline, C. (Ed.) (2018a) *Animals, Food, & Tourism*. Routledge.

Kline, C. (Ed.) (2018b) *Tourism Experiences and Animal Consumption: Contested Values, Morality and Ethics*. Routledge.

Markwell, K. (2015). Birds, beasts and tourists: human-animal relationships in tourism. In Markwell, K. (Ed.) *Animals and tourism: Understanding diverse relationships*, pp.1–23.

Matsunaka, K. & Koda, N. (2008). Acceptance of dog guides and daily stress levels of dog guide users and nonusers. *Journal of Visual Impairment & Blindness*, 102(5): 295–304.

McCabe, S. & Diekmann, A, (2015). The Rights to Tourism: Reflections on Social Tourism and Human Rights. Tourism Recreation Research, 40(2): 194–204.

McKercher, B. & Darcy, S. (2018). Re-conceptualizing barriers to travel by people with disabilities. *Tourism management perspectives*, 26(1): 59–66.

Mesquita, S. & Carneiro, M.J., (2016). Accessibility of European museums to visitors with visual impairments. *Disability & society*, 31(3): 373–388.

Michopoulou, E., Darcy, S., Ambrose, I., & Buhalis, D. (2015). Accessible tourism futures: the world we dream to live in and the opportunities we hope to have. *Journal of Tourism Futures*, 1(3): 179–188.

Nyanjom, J., Boxall, K., & Slaven, J., (2018). Towards inclusive tourism? Stakeholder collaboration in the development of accessible tourism. *Tourism Geographies*, 20(4): 675–697.

Palmer, C., & Andrews, H. (Eds.). (2019). *Tourism and Embodiment*. Routledge.

Poria, Y., Reichel, A., & Brandt, Y. (2011). Dimensions of hotel experience of people with disabilities: an exploratory study. *International Journal of Contemporary Hospitality Management*, 23(5): 571–591.

Price, H. (2017). The seven stages of life for a guide dog, from puppy to retiree. Available at: https://www.countrylife.co.uk/out-and-about/dogs/seven-stages-life-guide-dog-puppy-retiree-156677#lf60yzcHPPiGsmc3.99 Accessed: 19/11/19.

Richards, V., Pritchard, A., & Morgan, N. (2010). (Re) Envisioning tourism and visual impairment. *Annals of Tourism Research*, 37(4): 1097–1116.

Richards, V. (2013). (Re) Envisioning the tourism experiences of people with vision impairment (Doctoral dissertation, Cardiff Metropolitan University).

Rickly, J.M. (2018). Considering service animals in tourism. *Annals of Tourism Research*, 71(1): 57–58.

Ryan, C. (2000). Tourist experiences, phenomenographic analysis, post-postivism and neural network software. *International Journal of Tourism Research*, 2(2): 119–131.

Saunders, M., Lewis, P., & Thornhill, A. (2016). *Research methods for business students* (Seventh ed.). Harlow, England: Pearson Education Limited.

Small, J. (2015). Interconnecting mobilities on tour: Tourists with vision impairment partnered with sighted tourists. *Tourism Geographies*, 17(1): 76–90.

Small, J. & Darcy, S. (2010). Tourism, disability and mobility. In Cole, S. and Morgan, N. (Eds) *Tourism and inequality: Problems and prospects*, pp.1–21.

Small, J., & Darcy, S. (2011). Understanding tourist experience through embodiment: The contribution of critical tourism and disability studies. *Accessible tourism: Concepts and issues*, 72–96.

Small, J., Darcy, S., & Packer, T. (2012). The embodied tourist experiences of people with vision impairment: Management implications beyond the visual gaze. *Tourism Management*, 33(4): 941–950.

Swain, J., French, S., Barnes, C., & Thomas, C. (Eds.). (2013). *Disabling barriers-enabling environments*. Sage.
Tomkins, L.M., Thomson, P.C., & McGreevy, P.D. (2011). Behavioral and physiological predictors of guide dog success. *Journal of Veterinary Behavior: Clinical Applications and Research*, 6(3): 178–187.
UK. 2010. "Equality Act". UK Government, London, UK. [online]. Available: http://www.legislation.gov.uk/ukpga/2010/15/pdfs/ukpga_20100015_en.pdf Accessed on: 29/09/18
UK Government (2019). Pet Travel to Europe after Brexit [online]. Available: https://www.gov.uk/guidance/pet-travel-to-europe-after-brexit Accessed: 19/11/2019
United Nations General Assembly (1948) Universal Declaration of Human Rights in Higgins-Desbiolles, F., (2006). More than an "industry": The forgotten power of tourism as a social force. *Tourism management*, 27(6): 1192–1208.
UNWTO (2017). Tourism and the sustainable development goals–journey to 2030. [online] United Nations World Tourism Organisation. Available at: https://www.e-unwto.org/doi/pdf/10.18111/9789284419401 Accessed on: 01/10/19.
Urry, J. (1990). *The Tourist Gaze*. London: Sage.
Veitch, C. & Shaw, G. (2011). Disability legislation and empowerment of tourists with disability: The UK case. In Buhalis, D. and Darcy, S. (Eds.) *Accessible tourism: Concepts and issues*, pp.62–72.
Vila, T.D., Darcy, S., & González, E.A. (2015). Competing for the disability tourism market–a comparative exploration of the factors of accessible tourism competitiveness in Spain and Australia. *Tourism Management*, 47(2): 261–272.
World Health Organisation (2019). Disabilities. [online] Available at: https://www.who.int/topics/disabilities/en/ Accessed: 19/11/19.
World Health Organisation (2010). Global Data on Visual Impairment. [online] Available at: https://www.who.int/blindness/publications/globaldata/en/ Accessed: 19/11/19.
Young, J. & Carr, N. eds. (2018). *Domestic animals, humans, and leisure: Rights, welfare, and wellbeing*. New York, USA: Routledge.

Christina Vasilopoulou

10 The donkeys of Santorini: Workers or slaves?

Introduction

In a recent press conference, the Prime Minister of Greece stated that Santorini is one of the most touristed islands of the Aegean Sea. Santorini attracts more than 1.5 million tourists per year, with half of them arriving by cruise ships, equating to roughly 80,000 tourist arrivals daily (Spilanis, 2017, p. 14, 21). Hundreds of donkeys are used every summer for carrying these tourists between the old port to the town of Thira, a route that is also available via a cable car or on foot. The donkeys await the arrival of the cruise ships and climb hundreds of steps under extreme heat. In order to justify the exploitation of the donkeys, their legal owners argue in favor of tradition, i.e. that their use, along with mules and horses, has been taking place for thousands of years in many barren Aegean islands. They also argue that the architecture of the villages at the higher-altitude areas of the islands, where the narrow alleys are not accessible by cars, make the use of donkeys unavoidable even today.

Outside of the tourist season these animals continue their labor by carrying construction materials for the tourism establishments. In the heart of this exploitative process emerges the question of their position and subjectification in the biopolitical field of Santorini, and Greece in general. For thousands of years, the donkeys have been classified as "labor animals," a role that today is recognized by most of the Greek equine protection organizations. Modern donkeys are a hybrid, domesticated species that has been bred especially for human needs. However, whether these facts justify the deprivation of their right to freedom is highly contested.

In this chapter, I will examine the emerging questions of whether these animals are laborers, slaves, or mere machines, whether the quality of their relation with their owners is a matter of collaboration or of power, as well as how recent protests against the use of donkeys were constituted. The findings come from an ethnographic and a historical study, along with qualitative interviews, and information collected from the mass media. I will focus on the conflicting discourses that come from various perspectives, such as the donkeys' owners, local community and authorities, tourists, animal rights activists, and national and international media that cover the subject. I will also examine the theme in relation to the rise of mass tourism and the capitalist reconstruction of the island during the last decades, in particular, how this has led to the intensification of human and non-human labor and the alienation of the traditional way of life and values. Finally, I will explore the ethnological aspects of the donkey's role in the Greek and island folk culture, in relation to the use of donkeys in the tourism industry today. While there is

https://doi.org/10.1515/9783110664058-010

considerable research on the impact of tourism to the Greek people, the matter of the use of animals in the tourism industry has, as far as I know, not been researched. Thus, this research initiative will add a much needed non-human animal perspective to our understanding of tourism in the region.

Non-human animals in anthropogenic environment

The embodiment of non-human animals into the anthropogenic environment has been a long process; firstly, through the gradual conquest of the natural world by human population growth and, secondly, by the domestication of some non-human species and the creation of some new human-dependent species. The process of domestication has since led animals and humans into coexistence and has also constructed common narratives, despite the fact that the animals themselves remained invisible until nowadays and that their presence is preserved only through popular myths.

The use of domesticated animals is morally justified on an utilitarian basis, where they are considered as quasi-tools, used to serve human needs. In parallel with the study of their symbiotic relationship with humans, arguments regarding animal labor and questions on the equal involvement of human livelihoods have also emerged. Korsgaard (2018), in her attempt to answer the question of whether animals should be used as property or as subordinates, has set an example of their involvement in the most abhorrent practice of human warfare, concluding that since domesticated animals are now granted legal protection and rights, living within the borders of each state, their use or contribution is more justified than it is with the wild ones, especially the marine animals who live in “international” waters.

In addition to domestication as a process of genetic control of species that results in a more timid and disciplined behavior, individual animals undergo a taming process to obey orders. The discourse of animals as co-workers is introduced in Donna Haraway’s (2003) *The Companion Species Manifesto*, where the dog training process is described not as an act of subjugation of the animal, through the removal of free will, but as a symbiotic co-evolutionary practice where the domesticated animal finds its place next to the master willingly and does not wish to be another animal that experiences more freedom. However, according to Haraway, lazy or friendly shepherd’s dogs were killed to enrich the genetic pool with dogs more inclined to guard their herd and space (Haraway, 2003, p. 68). In this case, dogs were clearly treated as working animals whose abilities made them worth living, and their reward for the work was not just their nutrition and their accommodation, but allowing them to live, too. This utilitarian attitude towards domesticated animals makes this working condition more akin to slavery, as only in slave societies do we encounter the right to life and death by the master over the slave. The slave is perceived as an inferior, or as an object, and is in any case owned, just like an animal (Jacoby, 1994, p. 90).

The concept of work as an entirely human process was described by Engels. According to Engels and the contemporary Marxist theory, animals are unable to transform their environment, and the evolution from monkey to human, the "humanization" of the monkey, is rooted in the work process that also makes humans a dominant species (Engels, 1876). However, beyond human labor, some animals were forced to become producers themselves, as tools and machines; and through the process of domestication there was also a process of humanization. For example, the ox, the mule, the donkey, the goat, the horse, and the dog, among many others, left the "animal kingdom" and entered the human world, a world dominated by production and ownership. These domesticated animals go to the field on a daily basis; they dig, plow, carry, and guard the territory, just like humans, and they create a product through their labor using their abilities or qualities. For animals that are not intended to be slaughtered or to be used for milk production, we cannot easily say that they are treated as tools or machines, since their vital qualities render them useful and productive. They seem to be tools when they dig and carry, but when they have babies that are destined to become meat, they seem more like producing machines. Hence, the animals are a sort of *subalterns* where their vitality is recognized but the value of their life and self-substance is not. To describe the contribution of the animals to the transformation of the world, as well as the human societies' perception of it, Chhabra proposes the term "laboring bodies" (2019, p. 170).

Through their integration into human historiography, animal bodies transcend the limits of animality and are integrated into human societies as stakeholders. The divide between humans and animals in the *domesticity era* is very deep and is not disputed (Bulliet, 2012, p. 72) – a conception that is assisted by the Judeo-Christian tradition of Genesis, where God gives man sovereignty over the land, as this was reflecting the political and social reality of the time (Patterson, 2014, p. 34). The domesticated animals are perceived as an extension of the *oikos* (family) and very often have names, play with the children, and participate in the common life, but always as inferiors.

For modern urban people this relationship has changed. While a part of the population recognizes the need for the animals' rights to be respected on the basis of respect for life and freedom, on the other hand, the majority has moved so far from the animal condition that confronts it as inherently non-human or even non-living. However, there are also societies where traditional lifestyles are intertwined with modern ones, such as the islands of the Aegean, where people revive a folklore way of living, while at the same time setting up a modern capitalist structure to cope with the ever-increasing demand of its biocultural product. The animals that are part of this approach, including Santorini's donkeys, are experiencing a new form of exploitation as objects of sightseeing, while their common use of carrying people and heavy objects is increasing.

Research design

The island of Santorini

Santorini is located in the Southern Aegean Sea, in the island chain of Cyclades, and has been inhabited since antiquity. It is a volcanic island with a unique landscape, making it one of the most visited places in Greece. With the expansion of tourism in the Greek islands during the last five years (INSETE, 2020, p. 32), Santorini has also become a wedding destination and many luxury establishments have been built, causing serious damage to the environment (Sarantakoy & Terkenli, 2019, p. 418). The alienation of the landscape has been followed by an alienating effect on human relationships. Research shows that the island has witnessed overtourism (Sarantakoy & Terkenli, 2019, p. 42) and there are threats to the natural environment, due to the geological fragility and the limited natural resources of the island.

While the term overtourism has been used widely the last few years, Dodds and Butler (2019) suggest a number of limitations in the use of the term in order to avoid its potential misuse. Those preconditions refer to the competition for space and amenities between the locals and the tourists, to the increasing number of tourists, more affordable travel expenses, new groups of tourists arriving through time, dominance of the growth in authorities and the state of mind, short term focus of municipalities on tourism expansion, over-promotion of the destination by mass and social media, lack of control on tourists numbers and imbalance of power and fragmented tourism stakeholders (Dodds & Butler, 2019, p. 6). Unfortunately, Santorini meets all the above conditions. The number of tourists has expanded, in particular more tourists from Asian countries are coming every year, and the travel cost has been minimized due to direct international flights to the islands. Thousands of people crowd in Caldera to see its famous sunset. The number of tourists, especially of daily visitors from cruise ships, increased so much that the authorities were forced to put a daily limit on the numbers of the cruise passengers that are allowed to disembark (Jamieson & Jamieson, 2019, p. 220). However, the municipality and the local businesses are interested only in economic growth rates and are indifferent to or underestimate the problems that are caused in the everyday life of locals and in the landscape. The result of growth is a great housing problem even for the people that work in the tourism industry, as it is often obvious from the public doctors and teachers who refuse to work in the island, and from the problem of water deficit as well. During the summer season there is an all-day traffic jam between the port and the village of Fira. The tourism development that has attracted international investors in the hotel and other businesses has also resulted in an imbalance of power between the locals.

Additionally, overtourism has a great impact on the use of donkeys in multiple jobs within the tourism industry, giving a new perspective to the moral decline for the sake of the expanding profitability. The donkeys become a means of transport for people, rubbish, rubble, and building materials, while they remain an integral part of a local imaginary who wants them to be a fundamental element of a folk identity that has long disappeared. Hence, in this particular context the question of whether we treat animals as property or as a subordinate population emerges again.

Methodology

The present study follows a qualitative approach, through field observation during a visit on the island in July 2019, when most donkeys were recorded and photographed during the daytime in the Caldera area. In addition, four telephone interviews were held with three inhabitants of Santorini, plus an additional interview with an activist who went to Santorini in the summer of 2018 in order to organize a protest and to consult with the local authorities and the representatives of the donkey owners. The interviews were conducted using a semi-structured interview approach. The duration of each interview was one hour. Three were conducted in October 2019, while the last one that took place in April 2020.

My visit to Santorini in July 2019, during the high tourist period, was intended to be an aspect of auto-ethnography that proposes to see the situation from the perspective of a visitor. Of course, the view of the researcher should never be estimated as neutral, so I consider it as critical to situate myself as a researcher not simply participant-tourist in the study. This methodology has been proposed by Tilley-Lubbs (2016, p. 3) for researchers that do ethnography in dominant cultures. In my research the dominant culture of interest is the donkeys themselves. Moreover, while I clearly advocate for the abolition of their use for human purposes, this does not change the fact that for the animals I am a part of the ruling class of humans and that I had the privilege to visit the island in order to observe their painful lives.

In order to maintain anonymity, the names of the participants have been changed to other common Greek names. The activist Andreas, an elder Greek animal advocate activist who has organized several animal rights protests, had a central role, as through his personal experience on the island during his visits there, he has been connected with the local population and has witnessed the changes taking place in relation to animals. Additionally, he has delved into the muleteers' relationships with the animals. The study of the local community and the internal relations of the people who use the animals helps us to understand the issue more fully and to be able to propose effective solutions. We must not forget that individual relations are constituted locally and, therefore, should be considered as unique. The other three participants Kornilios, Anastasia, and Roussetos are local young people (24–36 age) who have been living in the island since their birth or childhood. They are all facing the impacts of overtourism and are aware of what goes on with the donkeys as they work near them, and sometimes they express publicly their opinion against the donkeys' exploitation.

The present research encountered two basic limitations: a) the difficulty to approach the animal owners themselves, due to an attack to an activist a few days before my visit (KeepTalkingGreece, 2019) and b) the inability to hold a discussion with the locals due to the unstoppable work patterns of the islanders. This was added to the difficulty of finding local people to talk about the matter, because of concerns about negative publicity stemming from the international media coverage of the donkeys use on the island during the last few years.

Findings

Visiting the island and observing the phenomenon

Santorini is a crescent-shaped volcanic island group; the main island is Thira, in the center there is the volcano, and to the northwest is the little island of Thirasia. The most famous villages of Thira are located in the inner line of the semicircle called Caldera. These are Firá, the island's capital, the village of Oia, and Imerovigli. Each of these villages are built on steep slopes and have been built with the assistance of donkeys to a great extent.

During my visit to the island in July 2019 the heat was unbearable. There were thousands of tourists and the island looked more commercial than I could imagine. I had visited Santorini as a little girl in 1998, and the first thing I remember were the donkeys carrying tourists and their luggage. Also, there were many donkeys in Caldera wearing heavy wooden saddles under the hot sun with no shelter (Figure 10.1). When arriving in 2019, I was surprised to see no donkeys in the main port, though the distance we traveled with the bus was over 20 minutes steep uphill. It feels unimaginable that an animal would be forced to walk this distance in the past. Today en route from the main port to Firá, one sees scattered equines tied up under the hot sun. Most of the Greek islands use donkeys for various jobs and some have a few of them in the port available for short rides, but only in Santorini is this phenomenon so pervasive. Inside the village are many souvenir shops, selling donkey-themed souvenirs. Stuffed dolls, statues, t-shirts, donkey milk soaps are everywhere, and as we move toward the Caldera the numbers increase.

It is five in the evening and a local is already pulling the donkeys out of the area where they lift tourists (Figure 10.2). He is wearing a blue t-shirt and a hat, and I assume that this is a kind of costume for the muleteers, because the other muleteers were wearing similar outfits in a photo taken in 1998 (Figure 10.3). He has nine animals in total, each tied with ropes to the one behind, while he is riding the first animal. A girl from a nearby shop gives him some sort of juice packed in a bottle. The animals look tired though they have no obvious sign of abuse. They all wear an iron muzzle in their mouth, which automatically brings to my mind images of

Figure 10.1: The donkeys in Caldera Summer 1998.
Source: Author.

slaves of the 1800s, a comparison also made by Marjorie Spiegel (1988). Certainly, these animals cannot drink water or bray. They Also cannot stop on the way as they are tied. I notice that these are not donkeys but mules, larger and stronger animals derived from breeding a donkey with a horse. The muleteer is a man over 50 years old wearing a worn-out hat and clothes, holding in his hand a rod from which hangs a rope with a metal object, a tool he uses to hit animals. The saddle and the ropes are very tight, and underneath the saddle there is a colored fabric as an example of traditional culture.

As we walk down the steps, the path that forms a trail of walking animals becomes visible. Some animals stop under the heat of the sun. As we approach them, the smell of feces and animals become unbearable. By the last turn before the first shelter, there is a big sign: "Equestrian Station – Entrance" in Greek and "Donkey Station – Entrance" in English (Figure 10.4). This difference in language unfolds a reality: Because this kind of work is more physically demanding, it is done by mules instead of donkeys. However, the donkey remains a registered trademark to attract tourists. This misappropriation of the donkeys against mules happens, arguably, due to the greater familiarity of donkeys through myths in popular culture. Moreover, the mule is not a pure species, as it cannot reproduce itself: the mule is a half-donkey, whose non-racial purity does not reinforce the narrative of the Santorini donkey's locality as a trademark (Clutton-Brock, 2012).

Figure 10.2: The muleteer with the animals, July 2019.
Source: Author.

According to Gregory (2007, p. 200), in Ancient Greece the donkey was hierarchically inferior to the horse and mule, because of its diminished capacity, and it was mainly used only by the poorest social classes to carry lighter loads, because of its endurance, patience, and its limited food needs. Further, Gregory argues that the mule's admiration in antiquity is based on the notion of work, and it is perceived as an artificial machine that increases productivity, which stands between the free spirit of the horse and the stubbornness of the donkey (2007, p. 208). However, according to Bulliet (2012, p. 195), the reason for the domestication of the non-horse equine was not riding them but having them carry objects. In fact, he suggests that the older saddles were intended for loads and not for humans. So while the horse is perceived as a companion whose intelligence and sensitivity are praised, the mule and the donkey are mere beasts of burden. Beyond its racial purity, the donkey provides the tourist with a traditional tale, as an animal used by the poor people of the islands for their survival, along with a good farmer living a

Figure 10.3: The muleteer and the donkeys, Summer 1998.
Source: Author.

common life on the margins of survival together with his animal. Therefore, the donkey is instrumental in perpetuating a folk myth about a long-lived symbiotic relationship between islanders and donkeys. The donkeys are not only integrated into the cultural landscape, but also into the physical landscape of the island.

A second rider with eleven animals follows. He is older than the first. His mules are not so tightly tied and do not wear a mantle, and they all have assorted trimmed yellow saddle substrates. The animals are adorned with elaborate necklaces and other fabric details, fully harmonized with the effort to highlight the island. As the animals are also touristic products, they are presented as beautifully as possible. At the beginning of the route there are some shelters and two informative signs. One is about the cost of the mule route (6 euro) and the other one about the additional cost of photographing on the mule, overlooking the sea (4 euro). About half of the animals that are tied up are in the shade, while the rest are very close to the rock under the hot sun, with the temperature over 40°C (104°F) and the volcanic soil intensifying the heat. However, few tourists are seen, as the cruise ships have not arrived yet. A tourist father carries his son on his back and he is taking a selfie, depicting the mule and the conductor.

Nearby is the cable car, built in the early 1980s with the endowment of ship owner E. Nomikos (Thira, Santorini Municipality, 2020). The profits of the cable car

Figure 10.4: The donkey station entrance July 2019.
Source: Author.

operation go to the municipality and the muleteers; however, it is now managed by a private company. In the main square of the island is a huge plush donkey and tourists are photographed with it (Figure 10.5). My attempt to reach the footpath on the Fira-Imerovigli route at that time was almost impossible due to the choking heat, and many animals had already gone back to shelters. The objective conditions in which animals "work" are adverse, and the animals have no way out of them.

The people's point of view

Through the interviews with young islanders who disagree with the exploitation of the donkeys, and the activist who traveled from his city to stop this practice, two contradictory narratives can be observed. The islanders' ideas were stricter and less hopeful regarding the animal owners while the activist felt there may be alternatives to the established relationship of the muleteers with the animals. These two perspectives help us to delve deeper into some key questions that arise

Figure 10.5: The donkey simulacrum, July 2019.
Source: Author.

when we look at the donkey phenomenon in Santorini, and at animals that are used as traditional tourist attractions in general. It has become not only conceivable but common for animals to be chained behind one another, like slaves, and go among the inhabitants and the tourists of this island, yet be unnoticed. What is it that allows animals to have this transparent presence in space? These issues arise in relation to the extreme touristic development of the island that, on the one hand, brings economic comfort to the local population, relieving them of the need to have such animals, and on the other hand allows them to transform their former companions into marketable machines.

Working conditions

The donkeys and mules of Santorini work from dawn to dusk. For these animals, their work in the service of the tourism industry does not finish with the uphill climb of tourists from the old port to Caldera. Based on what the two islanders, Kornilios and Anastasia, have said, two more routes have been added: a three-hour journey from Oia to Imerovigli and Fira, and another route from Oia to the small harbor, where excursions set off for the island of Thirassia, where, according to Kornilios, the path is a little more neat and the route is shorter. In addition, Anastasia pointed out that in the village of Pyrgos the donkeys are tied up for the whole summer period in a spot under the sun, waiting for tourists to take pictures of them, to take a stroll with them, or simply to carry their suitcases to the hotels.

Tourism has brought the need to create more and more lucrative lodging, most of which are located at the steep region of Caldera where it is not possible to transport building materials by mechanical means. The same situation appears across the island of Thirassia, where the donkeys are used for carrying building materials, as noted by activist Andreas who has gone to Santorini more than five times for the purpose of donkeys welfare. During the winter and summer periods, the equines carry ropes and building materials, participating in renovations or building constructions. At the same time, the tourism activity produces tons of rubbish from the hotels and the restaurants. The working day for the animals start at 5am, when the rubbish is loaded and taken to the garbage truck; this garbage collection is undertaken by a private company that pays the muleteers.

According to Anastasia,

> it's not the owners themselves that guide the animals, they are hiring other people, especially immigrants, maybe from Pakistan. I don't know how much they pay them. The owners only do the routes from the old port. Maybe they do so in order to give the impression that the animals in Caldera are not the same.

Similarly, according to Kornilios, they do so "because they want to collect the money themselves from the tourists [in Caldera]." Roussetos noted that many animals have died during the carrying, and several times he has seen donkeys and mules that cannot breathe due to the heavy carriage.

Kornilios and Anastasia estimated the number of animals to be about 200, but Andreas, the activist who negotiated with muleteers and the municipality in the summer of 2018 said that the animals were more than 500. Although the existing local laws permit up to 10 animals per owner, each muleteer had more than 15, and it is speculated that they possess many more undeclared. According to the activist Andreas, the last few years there was an attempt to electronically mark the animals, but the animal welfare group that was in charge of the project did not finish it.

The electronic marking (microchipping) could discourage the owners to stop abandoning or brutally kill the animals when they are not able to work. According to Anastasia, these animals are expendable for their owners, as "when they are too exhausted or too old they kill them or leave them outside the animal shelter on the island." The shelter is called Santorini Animal Welfare Association (SAWA) and hosts many old animals, including dogs, mules, and pigs.

The impact of tourism on animals

Tourists are on the island only for a few days; many do not feel responsible for what is happening, and they do not witness all these animal welfare violations on a daily basis. Additionally, for the majority of them, the animals are a part of the cultural landscape. Roussetos suggested that the tourists do not see the mistreatment during the garbage collection hours, which is the hardest part of the donkeys' work. Furthermore, he believes that the cruise passengers are not aware of the issue of animals or are unsure of how to change it. Relatedly, Tully and Carr (2020), in their article on donkey's oppression in the Blackpool tourist area in the UK, propose that tourists consume not only the experience of the walk but also paint the dominant ideology that legitimizes the exploitation of animals.

The inhabitants live in a frenzied pace of mass tourism and work from dusk till dawn or vice versa, just like the animals themselves. As Anastasia claimed, "The island's authorities argue that there are not enough police forces to oversee the working conditions of animals." At the same time, the working conditions for the non-local seasonal workers are described as dystopian. As Kornilios said:

> Even the rural doctors are often forced to stay inside the hospitals they work at because there are no available homes to rent, due to the fact that all the locals rent their residences to tourists for an extremely high price.

According to Kornilios:

> Apart from the animal matters, Santorini is a morally alienated place in general. People used to live in rivalry between them. Because of the big changes that tourist income has brought to the island, you can see a man having five hotels and himself living in a shack to even rent his house.

According to the local interviewees, real estate became a business for all. Kornilios, who is also a teacher, proclaims:

> The kids are in a tragic lack of culture. They are not interested in education. Parents just want them to keep the businesses. A large number of them are not educated; they are not interested in anything other than tourism and money. Every trace of humanity and compassion is gone.

This has also affected the potential for people who want to act on behalf of the animals. As he describes:

> The working pace is very fast and the young people have a hard time dealing with the issue of donkeys. There was once a group on Facebook called "Santorini donkeys," which was exposed the mistreatment of the animals, but the people didn't deal with it for a long time, and the donkey traders didn't seem to worry about.

The locals who were interviewed felt that is a sign that profitability is more important than other matter.

As for the owners of the animals, all four interviewees pointed out that these were family owned businesses. Kornilios described them as "organized tourism businesses," where owners make a lot of money by exploiting animals. Most of the money is undeclared, and he believes they seem to be among the richest people on the island as "they take 5 euro per person for a ride and the cruise ships disembark thousands of people every day." However, they constitute a small percentage of the island's labor force. Anastasia argued that they also have other businesses and do not rely on the donkey-oriented business exclusively. "They see it as a father-to-son tradition." Both argued that the mule owners use immigrant workers, but they work themselves or supervise them too.

On the other side is Andreas' view who visits the island as an activist in order to conduct on-the-spot inspections of the animals' stables. He was able to do the supervision after negotiations between the local authorities and animal rights activists. He described the muleteers as very poor people, who literally live together with their animals and their stables are an extension of their home. In financial terms, he believes that they do not make as much money as people believe and that some of them have to work in other islands during the winter. According to him, for the muleteers the tourist season is not long enough and the animals need care all year long.

The muleteers' relationship with the animals

There is a pervasive public perception of rural Greek people's lifestyles that the relationship of residents with their domestic animals is still much like that of 50–100 years ago. Today, whether it is farm animals or working animals, the intensification of animals' exploitation is firmly situated within false ideals about the past. Although a part of society has taken a moral view on non-human animals during the last few years, there still exists a reality that those who deal the commercial use of animals treat them as profit-making machines.

The relation of the muleteers to their animals cannot be differentiated from the relation of the dominant with the dominated. Despite the reciprocating work and cohabitation with humans, the animals themselves have not chosen to live there and, thus, they are constantly chained and subjected to the fear of whipping and beating. Roussetos said that the donkeys are very afraid of the muleteers and their pupils dilate from fear. Additionally, the muleteers use the male equines to control the female ones. As Roussetos described, the majority of the carrying animals are females, because it is believed that they are stronger and more patient, but the muleteer rides a male one in order to force the females to follow him. The aspect of gender influences the relationship of muleteers with the animals, as it is a process of double subordination: first through the category of the species, and then through the category of the gender.

Andreas, the activist, despite his clear opposition to any kind of animal use, managed to empathize with the muleteers and transcend the limits of the modern rational person. As a result, he was able to feel the compassion, the anxieties, and the "love" the muleteers have for their animals, although it is a love very different from what animal activists understand. Lydaki (2012) reminds us that transformations occur when tradition meets modernity; both humans and animals become hybrids of this change and enrich the world around us, concluding that the social sciences need to consider these relationships.

According to the three young locals, for the muleteers the animals are nothing but slaves. As Kornilios describes,

> these animals are only for work, neither encourage nor telling them bravo, the humane element in their behavior towards them completely disappears, on the contrary there is the brute-force.

The term "subjective investments," proposed by Porcher and Schmitt (2012, p. 45) for the qualities of cows, fits the case of the donkeys, but as Canavan critically declares, this point of view only acknowledges the subjugation to the human rule (2019, p. 45). Anastasia believes that the owners see the animals completely as

> slaves; there is no love, it's a sick relationship. [They see them] as mere objects. They adopt the narrative that "I am watering you, I am feeding you, you were born for this and no comment. As if they believe they are doing them a favor.

The speeding up and commercialization of tourism work has exacerbated the owners' treatment of the animals who, according to Anastasia, are now "making money out of it [which] makes [their attitude] even worse." She feels that they justify their actions toward the donkeys as they are specific creatures that were born to do this work, so they "use their anger and aggression on them." However, because the animals were never used to carry tourists and build hotels in earlier days, I feel the basic argument of tradition should be rejected. According to Kornilios, very few elder people communicate with the animals and have working relationships with them; additionally, he claims that they are the only ones who talk to the animals "naturally" as you would to a pet, all the other muleteers treat them as mere objects.

From the above discourse I wonder if in the consciousness of many young locals, the work of the equines in the tourism industry may not be considered a natural activity.

Also, the fact that the older people speak to their animals automatically transforms the donkeys from objects of use to subjects with the ability to understand rational communication or even respond to it. Thus, through a relationship characterized as natural, the predominantly non-rational animal, the relative of the horse (ά-λογον),[1] the mule, regains the ability to communicate and talk with the human being, who he lives with. But when someone is in possession of 10, 20, or even 50 animals, this personal relationship is alienated through the stress of daily work. Anastasia defines this relationship as problematic on its basis, as it always starts with usability and she typically says that "I don't know anyone in the island who has animals and loves them, except from the shelter owner."

Andreas, who met the muleteers during some official meetings and on-site inspections of the stables and the workplace, concluded that

> they are not bad [people]; they are illiterate people, and they do not realize that what they are doing is brutal to the animals. When I went to the stables I saw them literally living together with the animals, washing their clothes together with animals' sails, kissing them in the mouth, hugging them all day.

According to his observations, going up and down the Caldera together with tourists and the animals is a difficult task for the muleteers, which causes the muleteers to consider they are all a family where they work together in order to live; people and non-human animals, waking up from the dawn and sleeping late at night, spending their lives in the tourism industry. At the same time, in the discussions that took place in the town hall with activists' organizations and equine associations, the muleteers seemed willing to change their behavior towards animals, in order to avoid an impending ban on the use of donkeys. However, some of the animal owners have violently attacked activists from abroad and from other Greek cities, as well as locals who have tried to film them, in an attempt to defend their profession.

Animals in tradition transformed into a fetish for tourism attraction

According to Anastasia, tradition and a longing for the past are the main excuse for using the donkeys, and hence, the resulting animal mistreatment. The fact that the most picturesque places are built from the labor of donkeys and mules is regarded as a proof of the necessity to keep them (fira-santorini.com/santorini-donkeys, 2020). Although the animals are also used for construction works, this is not entirely apparent to visitors, who sees them simply standing in the sun and decides if the exposure in the hot sun is or is not a mistreatment based on their own standards. In our discussion,

1 In Greek language the meaning of the word for horse "άλογο" is "non rational," a being without logic or speech.

Andreas described the ride from the old port to Caldera as a "road to Calvary," where the animals are forced to do the trip 6–10 times per day, depending on tourist traffic. But the image of donkeys standing still or walking up the stairs with tourists on their backs, did not cause any kind of condemnation from the online community or the tourism magazines for the cruelty that was taking place. On the contrary, the worldwide spread of images of the Santorini's donkeys, both in print and in digital form, resulted in branding for the island. Although donkeys are used throughout the territory of Greece and in neighboring countries, when someone looks at these characteristic photos, they typically think of Greece and mainly of Santorini. The donkey has been transformed into a promotional product of the island and this use was never challenged. Additionally, much of the touristic infrastructure took the form of this animal that was projected on souvenirs of all kinds.

The animal advocates discourse

In the recent years, there has been a fierce response on the matter and the abusive use of animals came under harsh criticism and controversy within the island, as well as nationwide and globally. International welfare organizations like PETA, the Greek equine and animal welfare organizations (ESPI, donkeyland, Hippothesis), and several animal rights groups and independent activists have led campaigns in order to encourage the public to empathize with the donkeys. The awareness of the island itself and of the donkeys has led to a reverse of the welfare standards and has caused strong reactions, most of them aiming at immediate improvement of the donkeys' living and working conditions, and at the progressive abolition of animal use. In the summer of 2018, July 27–29, activists had planned a protest on the island and had arranged meetings with people from the municipality and with the animal owners too. The main demands, as articulated publicly, were the improvement of the working conditions of the animals, their health care, and the microchipping. All the stakeholders agreed on the above welfare standards.

The equine protection organizations (Donkeyland, Ippothesis, ESPI), in a joint statement, recognized the mules and the donkeys as working animals, thereby validating the muleteer's perception that these animals were born to serve human needs (gaidourohora.gr, 2018). This essentialist conception of the nature of the domesticated animals is rejected by the cases of individual animals who managed to survive without human input; one such example is the wild donkeys of Cyprus, who were abandoned by their owners after the 1974 invasion (Hamrick, Pirgalioglu, Gunduz, & Carroll, 2005, p. 108), and today prosper in the area while being a tourist attraction. By the end of the summer of 2018, few of the promises of the animal owners and the municipality were kept. But the impact of the demonstrations, in particular the defamation of the muleteers after beating the activists on the Caldera steps, led to a reduction in the use of donkeys by the tourists.

Conclusion: A plea for the future

One of the main arguments of the animal owners is that if the animals do not do this job, four to five people will be needed instead of each animal to do the heavy work, since for tourists there have been running the cable car for 40 years. Although the muleteers receive a percentage of the revenue from the cable car as compensation for the loss of their operation, they are unable to think of any activity outside the donkey exploitation. The activists suggested the creation of a park where animals can enjoy care and be visited by tourists, who would pay a symbolic fee to the owners. Roussetos proposed a paradigm change, in which "we should integrate donkeys in the island landscape as companion animals, to treat them with tenderness and admire them as sacred beings, as they do in India with the cows." Obviously, this would not be possible without a ban on their use as a means of transport. But the change has already begun, as the cruise lines and passenger boats are suggesting to the visitors to avoid the use of donkeys, while the issue is being discussed internationally too. The society of Santorini has achieved an increase in gross income in the last 20 years from the tourism industry, so they are not dependent on the animals to have a living. Donkeys still exist in order to fulfill the traditional image and could still exist in another non exploitative way.

This debate is dealing with the recognition of the identity and the rights of animals. Even the microchipping of the animals operates as a tool of recognizing them as a subjective entity, and could be the beginning of the recognition of agency and citizenship in non-human animals, as described by Donaldson and Kymlicka in their work *Zoopolis* (Donaldson & Kymlicka, 2011, p. 155). The animals can claim their entities in these marginal naturecultures where their relation to humans still exists and where they have not yet disappeared practically or spiritually from people's lives. Thus, the practices towards the equine animals of Santorini could evolve from an exploitative attitude to a new paradigm of respect and co-existence, where animals would be transformed from beasts of burden to subjects of freedom.

References

Announcement of the Equine Organizations for Santorini's Donkeys (July 27, 2018): http://gaidourohora.gr/%CE%B3%CE%AF%CE%BD%CE%B5-%CE%B7-%CF%86%CF%89%CE%BD%CE%AE-%CF%84%CE%BF%CF%85%CF%82/

Bulliet, R. (2012). *Hunters, Herders, and Hamburgers: The Past and Future of Human-Animal Relationships*. Athens: Editions of the 21st Century.

Canavan, J. (2017). "Happy Cow" Welfarist Ideology and the Swedish "Milk Crisis": A Crisis of Romanticized Oppression. In: Nibert David (ed.) *Animal Oppression and Capitalism, Vol. 1: The Oppression of Nonhuman Animals as Sources of Food*. Praeger, California.

Chhabra, H. (2019). Animal labourers and the law in colonial India, *South Asia Research, Vol 39 (2)*, 166–183.

Clutton-Brock, J. (2012). *Animals as domesticates a world view through history. The animal turn series. Michigan State University Press.*

Dodds, R. & Butler, W.R. (2019). The enablers of overtourism. In R. Dodds and R. W. Butler (Eds.) *Overtourism: Issues, Realities and Solutions* (pp. 6–22). DeGruyter.

Donaldson, S. & Kymlicka W. (2011). *Zoopolis, A political theory of animal rights.* Oxford: Oxford University press.

Engels, F. (1876). *The Part played by Labour in the Transition from Ape to Man.* retrieved from https://www.marxists.org/archive/marx/works/1876/part-played-labour/.

Gregory, J. (2007). Donkeys and the Equine Hierarchy in Archaic Greek Literature, *The Classical Journal, Vol. 102, No. 3* (Feb.–Mar., 2007), 193–212. The Classical Association of the Middle West and South, Inc. (CAMWS), retrieved from https://www.jstor.org/stable/30037985 01-10-2019 10:40 UTC.

Hamrick, R.G., Pirgalioglu T., Gunduz S., & Carroll J.P. (2005). Feral donkey Equus asinus populations on the Karpaz peninsula, Cyprus, *European Journal of Wildlife Research, Volume 51, Issue 2, 108–116.*

Haraway, D. (2003). *The Companion Species Manifesto, Dogs, People and Significant Otherness.* Chicago: Prickly Paradigm Press.

INSETE (Institute of Hellenic Tourist Businesses Association) (2020). *Tourism in Greece and in Main Antagonistic Destinations (2014–2018)*, Retrieved May 30, 2020 from https://insete.gr/wp-content/uploads/2020/04/Greece-and-Competition-2014-2018.pdf.

Jacoby, K. (1994). Slaves by nature? Domestic animals and human slaves, *Slavery and Abolition, 15:1, 89–99.*

Jamieson, W. & Jamieson M. (2019). Managing Overtourism at the Municipal / Destination Level. In R. Dodds & R.W. Butler (Eds.) *Overtourism: Issues, Realities and Solutions.* DeGruyter.

Keep Talking Greece (2019). Activist attacked and threatened while filming donkeys abuse on Santorini. Retrieved October 15, 2019, from https://www.keeptalkinggreece.com/2019/07/17/activist-attacked-donkeys-santorini-nfa/.

Korsgaard. M. C. (2018). *Fellow Creatures Our Obligations to Other Animals.* Oxford: Oxford University Press.

Lydaki, A. (2012). Geometry or poetry? Issues of methods and techniques in regional research. *Quality & Quantity, 46(1)*, 291–301.

Patterson, C. (2014). *Eternal Treblinka: Our Treatment of Animals and the Holocaust.* Athens: Green Institute.

Porcher, J. & Schmitt T. (2012). Dairy Cows: Workers in the Shadows? *Society & Animals, 20*, 39–60.

Santorini Animal Welfare Association (SAWA): www.sawasantorini.com.

Santorini donkeys. Retrieved July 30, 2020 from http://www.fira-santorini.com/santorini-donkeys.html.

Sarantakou, E. & Terkenli S. T. (2019). Non-Institutionalized Forms of Tourism Accommodation and Overtourism Impacts on the Landscape: The Case of Santorini, Greece. *Tourism Planning & Development, 16(4)*, 411–433.

Spilanis, G. (ed.) (2017). Santorini Tourist Observatory: Impression of the tourist situation activity and its effects on destination, SWOT analysis and alternative scenarios policy. *Local and Island Development Laboratory, Tourist Study and Research Laboratory.* Special Research Account Aegean University.

Spiegel, M. (1988). *The Dreaded Comparison Human and Animal Slave.* Heretic books.

Thira (Santorini) Municipality (2020) The cable car. Retrieved October 20, 2020 from https://www.santorini.gr/index.php?option=com_content&view=article&id=75&Itemid=62&lang=en.

Tilley-Lubbs, G.A. (2016), Critical Autoethnography and the Vulnerable Self as Researcher. In Tilley-Lubbs G.A., Calva S.B. (Eds.) *Re-Telling Our Stories. Imagination and Praxis (Criticality and Creativity in Education and Educational Research)*. Rotterdam: SensePublishers.
Tully, P.A.G. & Carr, N. (2020) The Oppression of Donkeys in Seaside Tourism. *Int J Sociol Leis* 3, 53–70.

Part 4: **Hidden labor**

Siobhan I. Speiran

11 Monkey see, monkey do: The work of primates in Costa Rican sanctuaries

> *Primates existing at the boundaries of so many hopes and interests are wonderful subjects with whom to explore the permeability of walls, the reconstitution of boundaries, the distaste for endless socially enforced dualisms.*
> – *Donna J. Haraway, Primate Visions*

Introduction

"The monkeys should not be here," I was told by a staff member at a wildlife sanctuary in Costa Rica. Though said in passing, upon reflection this aphorism neatly identifies the complex entanglement of the human and more-than-human bodies involved in non-human primate conservation.

With 75% of primate species (henceforth "primates") declining globally, including 36% of neotropical species (Estrada et al., 2017), the need for effective primate conservation grows more urgent. In a Costa Rican sanctuary, one may encounter four indigenous, neotropical primates: the Panamanian white-faced capuchin monkey (*Cebus imitator*), the black mantled howler monkey *(Alouatta palliata*), black-capped squirrel monkey (*Saimiri oerstedii*), and Geoffroy's spider monkey (*Ateles geoffroyi*). Each species is susceptible to a variety of anthropogenic pressures depending on their range and behavior. Based on my archival research and interviews with key informants, howler and squirrel monkeys appear most at risk of electrocution, while capuchins are at risk of direct conflict with humans, and spider monkeys are the most endangered due to habitat fragmentation and hunting. Monkeys are rescued and brought to sanctuaries for a number of reasons: electrocution, road collisions, conflict with humans and non-humans, orphaned, or confiscated as pets from homes. Typically located near the areas in which animals are rescued from and released, sanctuaries and rescue centers are at the forefront of this "interface of extinction" (Parreñas, 2018, p. 26) both literally *and* figuratively.

Costa Rica is a leading ecotourism destination, and presently its wildlife sanctuaries offer the only (legal) tourism attraction to encounter native wildlife up close beyond visiting natural reserves. The Ley de Conservación de la Vida Silvestre N°7317 (Wildlife Conservation Law), defines a rescue center as a site of wildlife management "whose objective is to rehabilitate wildlife that has been rescued, confiscated, or voluntarily surrendered for recovery and reinsertion into the natural environment when appropriate [. . .] they are not for profit and not open to the public." Conversely, it defines "zoo" as a site "that keeps wildlife in captivity [. . .] guarantees them adequate living conditions in an attractive and didactic way for the public. Its main objectives are the conservation, education, research, and exhibition of wildlife in a

https://doi.org/10.1515/9783110664058-011

scientific way." Such operations tend to avoid advertising as "zoos," however, as "sanctuary" better represents their values. Therefore, in this chapter I use that term. These two operations often go hand-in-hand; 5 of the 8 sites I visited are both zoos/ sanctuaries *and* rescue centers, meaning they housed both rehabilitant (i.e., slated for release) and permanent resident monkeys in distinct locations on the same site.

While the work performed at these sites is similar to that of a zoo – education, entertainment, tourism, conservation, etc. – these labors are recast in light of the sanctuary's position at this interface. With no government funding, and reliant on volunteer and tourism dollars to operate, sanctuaries are the last chance for injured monkeys to return to the wild. This is no easy task – some sites have as many as 10–20 rehabilitant monkeys arriving per month, while simultaneously caring for numerous permanent resident monkeys for life. Despite being the last vestige of hope for many animals, there is no published scholarship on animal labor at sanctuaries in Costa Rica to my knowledge.

On a global scale, the expanding market for ecotourism suggests a growing conscientiousness of sustainability amongst tourists. Conservation of species, ecosystem, and local culture are the foundational aims of ecotourism – a subset of wildlife tourism with roots in sustainable development theory (Wallace & Pierce, 1996). Ecotourism's ability to deliver on this triumvirate is not yet fully understood. In the absence of global regulation, ecotourism has been criticized as facilitating the neoliberalization of nature (Duffy, 2008) and commodification of animal bodies (Collard, 2014). More exploration of the diverse lives and experiences of wild animals in ecotourism, upon whom the industry depends, is needed. I posit that any operation which fails to prioritize the welfare and conservation of its animals cannot be considered sustainable (Carder et al., 2018; D'Cruze et al., 2017).

Fortunately, research suggests that sanctuary tourism – as a potential form of ecotourism–offers ethical animal encounters with good outcomes for the welfare and conservation of the species in their care (Moorhouse et al., 2015). The work of a sanctuary is to decommodify (Collard, 2014) through processes of rescue, rehabilitation, and release, and by providing good lives for those who are too injured or habituated to humans to return to the wild. Sanctuaries participate in environmental education by offering tours and outreach with local communities, as well. The growing popularity of sanctuaries as tourism attractions (Compton, 2019) tracks the increasing attention they have been receiving from academics (Alexander, 2000; Das & Chatterjee, 2015; Sebele, 2010). Nevertheless, there is still a lack of scholarship on non-human perspectives in this context. Integrating the interests and perspectives of animals in decision-making spaces where they have long been ignored (Ramp & Bekoff, 2015) is critical for wildlife tourism and conservation.

I attempt to "bring animals in" (Wolch & Emel, 1995) by examining the labor they perform in an industry that tends to assign them instrumental value and sideline their interests. Specifically, I focus on the lives of monkeys in Costa Rican sanctuaries, places where they are acknowledged as individuals with names and life histories. I bring attention to their welfare at these sites as it informs the (often invisible) *emotional, educational, entertainment,* and *ecological* labor they perform *with* and *for* each other and humans. I argue that assessments of primate sanctuary tourism as ethical and sustainable must consider the concept of monkey labor so that we can adequately speak about positive approaches to ecotourism.

This chapter speaks to a number of different disciplines, including: animal labor (Coutler, 2015), tourism studies (Fennell, 2012), animal geography (Hovorka, 2017, 2018a, 2018b), compassionate conservation (Paquet & Darimont, 2010; Ramp & Bekoff, 2015), and ethnoprimatology (Dore, et al. 2017). In particular, ethnoprimatology views "human and nonhuman primates as co-participants in a rapidly escalating realm of ecological and cultural change" (Fuentes, 2002, p. 1). Indeed, "the driving force behind the mixed-methods ethnoprimatological endeavor is to effectively conserve nonhuman primates" (Dore, 2018, p. 918). Considering how my research integrates both biological and social science methods to understand the complexities of human-primate relations, I aim to contribute to this field.

I begin this chapter with a discussion of the conceptual frameworks in which my research approach is rooted. I rely upon two key texts, the former a book on animal labor and the latter an intersectional perspective on primate rehabilitation: Kendra Coulter's *Animals, Work, and the Promise of Interspecies Solidarity* (2015), and Juno Parreñas' *Decolonizing Extinction: The Work of Care in Orangutan Rehabilitation* (2018). Next, I consider the dual roles of *resident* and *rehabilitant* that primates inhabit in this context. I then theorize on four types of monkey work including: *educational labor,* as spokesmonkeys for their species, *entertainment labor,* as living tourism artifacts, and *environmental labor,* as ecosystem servicers. The fourth, *emotional labor,* I posit underlies all the other forms of monkey labor, since the work performed by both monkeys and humans in this setting is ultimately a work of care.

Animal labor

Labor scholar Coulter offers the concepts of "interspecies solidarity" and "humane jobs" as a basis for improving the lives of working animals. She considers how work done *with, for* and *by* animals is a robust area for criticism that has yet to be fully explored in the labor and animal studies disciplines. She recommends workforces move away from "damaging patterns and towards more sustainable and positive practices and employment sectors" ("Humane Jobs," n.d.). While the work of tourism animals is not a major focus of the text, my chapter shares this call for "sustainable and positive practices" in the context of wildlife tourism – which I argue can be moved towards if we consider animals as participants in the industry. This logical step is easier for "obvious" animal work (i.e., a donkey carrying luggage, a circus elephant, etc.), but is conceptually challenging when tourism animals appear – at least, at first glance – to be passive recipients of the tourist gaze (i.e., a monkey in a zoo). Nevertheless, I argue that the work monkeys perform in the sanctuary context is indeed *work* – as invisible as it may be.

Monkey work supports human labor at this interface, reaching tentactularly into different realms critical to the function and sustainability of the sanctuary operation. A sanctuary may be the perfect mesocosm to explore what Coulter calls her "anifesto" for "interspecies solidarity" (2015, p. 150). This is "an idea, a goal, a process, an ethical commitment, and a political project. This concept of solidarity is underscored by ideas of empathy" (p. 150). Coulter recognizes the interrelation of "the emotional" and "the rational" in animal work and presents interspecies solidarity as "an invitation to broaden how labor as both a daily process and a political relationship are understood and approached" (2015, p. 151).

This chapter references and remixes Coulter's five categories of animal labor. These include: (1) *care work* is the provision of physical and/or emotional care, (2) *communication work* is the creation of a shared language of verbal and nonverbal cues, including unintentional signaling of emotional states through physical reactions, (3) *emotion work* includes how animals control and manage their emotional states, as well as the emotional connection between animals and the humans they work for/with, (4) *body work* refers to "how animals' bodies and physical abilities are used work" (2015, p. 73), and (5) *dirty work* is typically work "deemed degrading and/or undesirable" and "physically unpleasant" such as that which deals with waste, "bodily fluids" and "excrement" (2015, p. 32).

To conceptualize monkey work in sanctuary tourism, I use different categories of labor from, but based on, Coulter's categories. The *emotion* work I describe is the same definition, however, and underscores virtually all other types of labor. I conceived the categories of *entertainment, educational,* and *ecological* work during the early stages of fieldwork while contemplating the role of sanctuaries as sites of animal care, conservation, and tourism. A sanctuary cannot function in these realms without primates' active participation – not just passive cooperation – in these labors. Put simply, monkeys do irreplaceable work in sanctuaries – by the very nature of their monkeyness – that we cannot do ourselves.

Decolonizing works of care

The categories I used to classify monkey work speak to broader political economic cultural colonial processes, as described in *Decolonizing Extinction* (2018). Costa Rica, where I carried out my research, is a developing country in Central America where there is a history of neocolonial conservation (Langholz & Lassoie, 2001), meaning a "tendency to blame powerless local people for ecological change and to instigate or support unjust regimes of fortress conservation" (Lorimer, 2012, citing Brockington, 2002). Costa Rica is also a location where ecotourism tends to "reproduce preexisting patterns of [gender and landholding] stratification," such that men typically hold the highest-paying jobs in ecotourism, and there is a history of poor extractive laborers, such as miners, being forcibly removed from national parks (Horton, 2009, p. 93).

Approaching the work of care in two orangutan sanctuaries, Parreñas recognizes primates as "colonized subjects" (2018, p.9), such that orangutans have had their autonomy "arrested" as they are "constrained in a space shaped by colonial interventions on the land" (2018, p. 23). Parreñas' call to "decolonize extinction" requires a "fundamental reorientation toward others, especially nonhuman others, in which we accept the risk of living together, even when others' lives pose dangers to our own" (2018, p. 27–28).

Rescue centers and sanctuaries in the developing world may be considered decolonial projects because the humans and non-humans involved in them share the risk of causing each other illness and injury due to the proximity required to provide daily care. This is evident in Parreñas' accounts of the precarity of rehabilitating semi-wild orangutans in Indonesia, as well as my own similar findings in the Costa Rican context – two circumstances in which the rehabilitation work is occurring in the endangered primates' home countries. These areas historically – and perhaps continually – are shaped by "colonial interventions on the land."

A wildlife sanctuary is an intriguing phenomenon to "think with" in the case of animal work in ecotourism because, as Parreñas recognizes, it is "different from any other site in the world, as it allows for the layperson's direct and embodied experience of what is to be at the interface of species loss and vulnerability" (2018, p. 19). Parreñas' text compliments Coulter's because, while labor is not her central lens, Parreñas contemplates the work of care and conservation performed by *both* humans and primates in sanctuary settings, which requires "embracing the risk of vulnerability" (p. 27) for both parties. Parreñas identifies how the work of care in a sanctuary is "not necessarily affection," rather "a concern about the treatment and welfare of others" (Parreñas, 2018, p. 6). Later in this chapter, I propose that monkeys engage in emotion work (incorporating Coulter's body and communication work) when they are concerned with the welfare of their conspecifics, evidenced by vigilance, territoriality, and affiliative behaviors. I also consider the entanglement of education and entertainment work performed by sanctuary monkeys, which speaks to this experience of embodied tourist learning at these sites.

Though touted as a form of ethical, alternative tourism, unfortunately not all sanctuaries put animals first as one would expect (Cohen, 2013). It is imperative to explore this niche tourism phenomenon from non-human perspectives in order to separate the truly ecocentric organizations from those which have merely been greenwashed (Moorhouse, D'Cruze, & Macdonald, 2017). Understanding the work primates do in these settings is important to fully appreciate how conservation labor is, at its heart, a collaborative process; requiring cooperation between humans and non-humans. For example, a rehabilitator may raise an orphaned monkey from infancy to be released, but if a wild troupe does not accept the orphan, they cannot be released. Likewise, sanctuaries place multiple, unrelated monkey individuals into an enclosure with the hopes of forming a captive troupe, but if the individuals do not perform behaviors appropriate to their social rank to create a cohesive family, there is a risk of conflict.

Improving the conservation or welfare of sanctuary primates is challenging for all parties, and given that we are in the midst of the 6th largest extinction crisis in history, the repercussions of declining wild monkey populations have an impact on us as well. By neglecting to recognize that conservation work is a collaborative effort between human and non-humans, one cannot understand the complex entanglement of bodies, emotions, and vulnerabilities at this important interface. As those who have spent a significant amount of time performing embodied (particularly physical) labor *with* and *for* rehabilitant wild animals will likely agree (Birke, Hosey, & Melfi, 2019), it is challenging to regard animals as anything less than individuals, with personalities, preferences, life histories, and traumas. Sanctuaries cannot be truly sustainable, ecocentric spaces without interspecies co-working – and solidarity – which in turn requires trust; accepting the mutual vulnerabilities that we share with primates, and that they share with each other.

Methods

The ideas I present in this chapter are informed by my experience conducting fieldwork in Costa Rica. My methods are interdisciplinary, drawing on animal welfare science, ethology, animal geography, conservation, and tourism studies. I conducted fieldwork in Costa Rica during which I visited eight sanctuaries and rescue centers between June and August 2019 across four conservation regions: Área de Conservación Guanacaste, Tempisque, Osa, Pacífico Central, and Arenal Huétar Norte (Figure 11.1). Each region has a regional office to which the operation must submit trimesterly reports through their governmental conduit, an accredited biologist called a *regente*, on the status of the animals in their care.

The three key sites I visited have other wild animals in their care, mostly mammals and birds, but my survey of tourists demonstrated that a close-encounter with jungle-dwelling primates (and sloths) are the major draw for tourists to visit a sanctuary. At each center I interviewed the owners, staff, and local conservation authorities (n = 39). I remained at three key sites between 3–6 weeks to employ more in-depth methodologies. This included: observing the behavior of resident capuchins during tours; reviewing all primate care records; shadowing staff on clinical examinations, tours, and releases; interviewing Costa Rican people (n = 25) living near the sites; and surveying tourists (n = 88) after their tour.

Figure 11.1: Areas of Conservation Map.
Note: I visited regions 1, 2, 6, 9, and 10.
Source: SINAC (2018). Used with permission.

Site 1 is the only operation I visited that is for-profit and is also the largest and longest-running of the three, offering two daily tours six days a week. It is owned and staffed by Costa Ricans, has 20 resident primates, and also receives a few primates to rehabilitate and release each year (Figure 11.2).

Site 2 has more informal origins; an expat couple began receiving and rehabilitating wildlife nearly 20 years ago at their hotel. Eventually, they founded an NGO in a more isolated location and made wild animal care their full-time vocation. They are staffed by both Costa Ricans and foreigners but rely on (paying) volunteers for the majority of their labor and funding. They have 15 resident primates in the sanctuary and offer one daily tour, six days a week (Figure 11.3).

Figure 11.2: Site 1 (S1).
Source: Site 1. Used with permission.

Site 3 has been operating for approximately the past six years, founded when another sanctuary was closing down. A generous benefactor who owned a hotel in the area donated his land to build a sanctuary and rescue center for the suddenly homeless animals. Similar to Site 2, this operation has a mix of Costa Rican and foreign staff and relies on volunteers for most labor and funding. They offer four daily tours six days a week and have only six resident primates in the sanctuary (Figure 11.4).

Being monkey in Costa Rica: Resident versus rehabilitant

One of the most common maxims I heard during my research was "monkey see, monkey do." This phrase popped up especially on sanctuary tours. It was often said, jokingly, to account for the behavior of the Panamanian white-face capuchin monkeys (*cebus imitator*). Capuchins are known for their behavioral flexibility (McKinney, 2010),

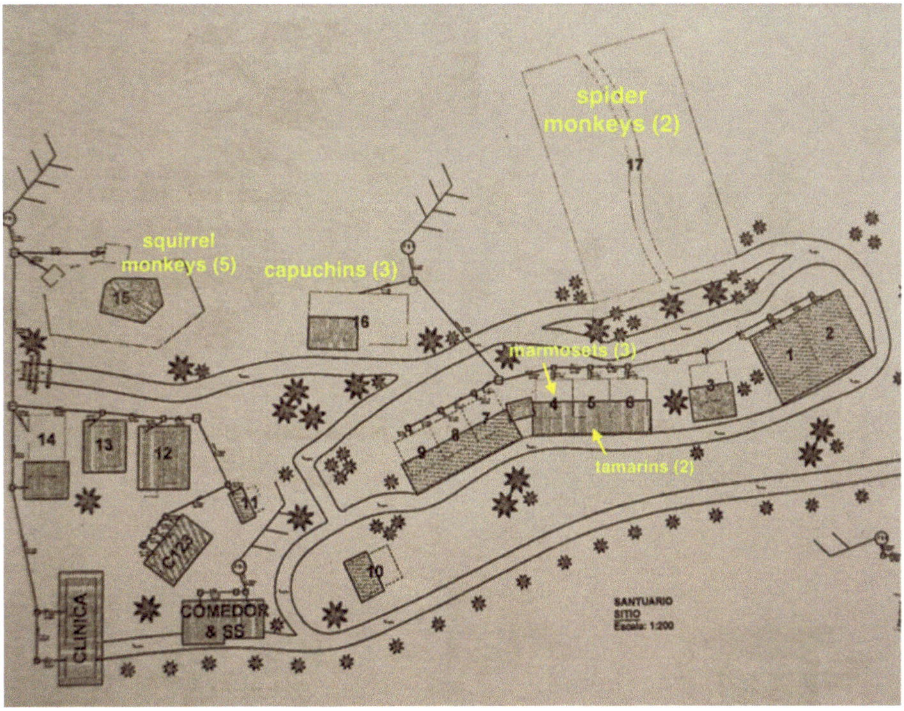

Figure 11.3: Site 2 (S2).
Source: Site 2. Used with permission.

a trait highlighted by tour guides. Ironically, this maxim is misleading; while capuchins' exploratory activities may be influenced by others in their troupe, primatologists purport they do not learn by imitating (Fragazy, Visalberghi, & Fedigan, 2004). Rather, capuchins become attracted to, or fearful of, items or situations based on how others around them behave (2004). At Site 3, I overheard a tour guide recount this story:

> There's a story I love to tell about the capuchins, because it shows just how smart they are. One day the volunteers were cleaning the shift cage and accidentally left one of the scrub brushes inside. Well, we saw one of the monkeys take the brush and start scrubbing just like the volunteers were doing. You've all heard of "monkey see, monkey do, right?"
>
> (Pers. Obs., Aug 2019)

This story was always met with laughs from tourists. The scrub brush in question is still in the enclosure, and on one occasion I did see a monkey scrubbing the perimeter with it. I also saw the same monkey pick it up, run bipedally and launch it at a tour group. It hit the fencing – this was also met with laughter.

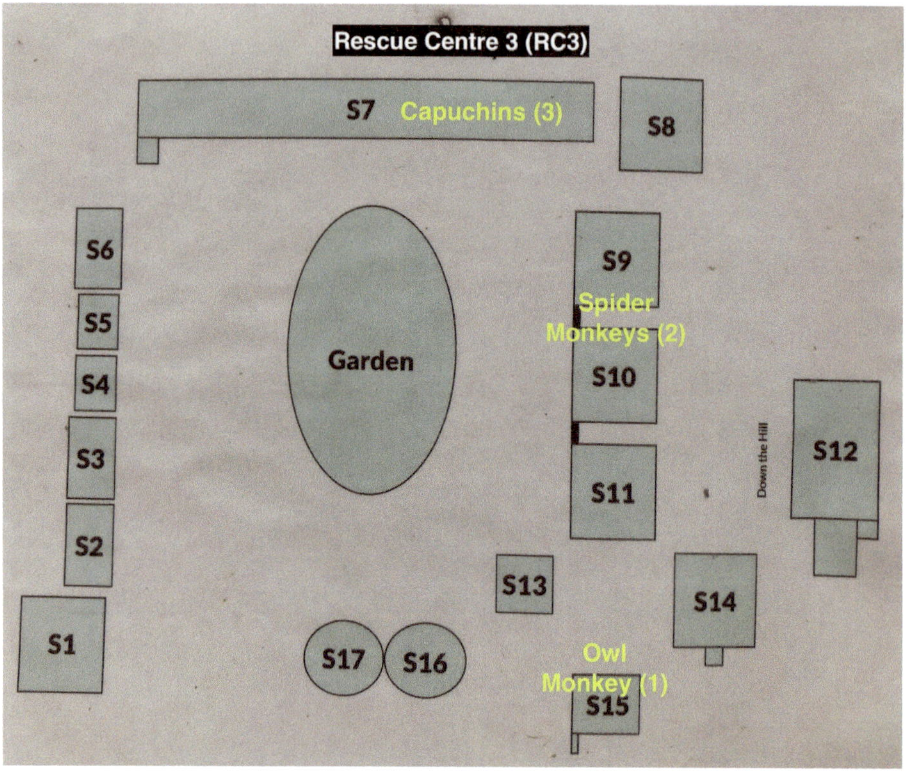

Figure 11.4: Site 3 (S3).
Source: Site 3. Used with permission.

On a global level, capuchins have a reputation as entertainment animals, performing *body work* and *communication work* as "organ grinders" and film and circus stars. Given their intelligence, capuchins have been employed as service animals, though not without controversy. In this role, they perform *care, communication, emotion, body,* and perhaps even *dirty work* as support animals for adults with spinal cord injuries and mobility impairments (Birke, Hosey, & Melfi, 2019). Capuchins are also in-demand as a wild pet (https://primatestore.com/forsale.asp). The YouTube channel, *MonkeyBoo*, follows the life of a pet black-capped capuchin and boasts over 2.1 million followers, and whose most popular video has 262 million views. There is evidence that sharing images of primates in domestic settings on social media influences the demand for them as pets (Lenzi, Speiran, & Grasso, 2020). These are examples of how capuchins have been known to perform all five types of labors defined by Coulter (2015) in various settings. Squirrel, spider, and howler monkeys feature less commonly in the aforementioned roles, but all are still victims of the illegal wildlife trade.

The following sections will explore the *entertaining, educational, ecological,* and *emotional* labor performed by primates in the Costa Rican context. My three study sites were operationally divided into a sanctuary and rescue center, each requiring its own license. *Sanctuary* monkeys, or *residents,* are those who live in captivity for life. The resident monkeys at the three sites were formerly pets or entertainers, therefore non-releasable. Rather, they act as educational ambassadors for tour groups. *Rescue center* monkeys, or *rehabilitants,* typically have not come from a previous state of captivity, and are unhabituated enough that they may be released – thus, they might not be seen by the public. The rehabilitants I observed were mostly victims of electrocution, road accidents, or were found ill or injured (Figure 11.5). Figure 11.5 was created by reviewing archival data related to the reasons for monkey intake, outcomes for monkeys (euthanasia, release, etc.), and species trends. This data is comprised of records from Sites 1 (n = 22 records), 2 (n = 185), 3 (n = 91), and a site

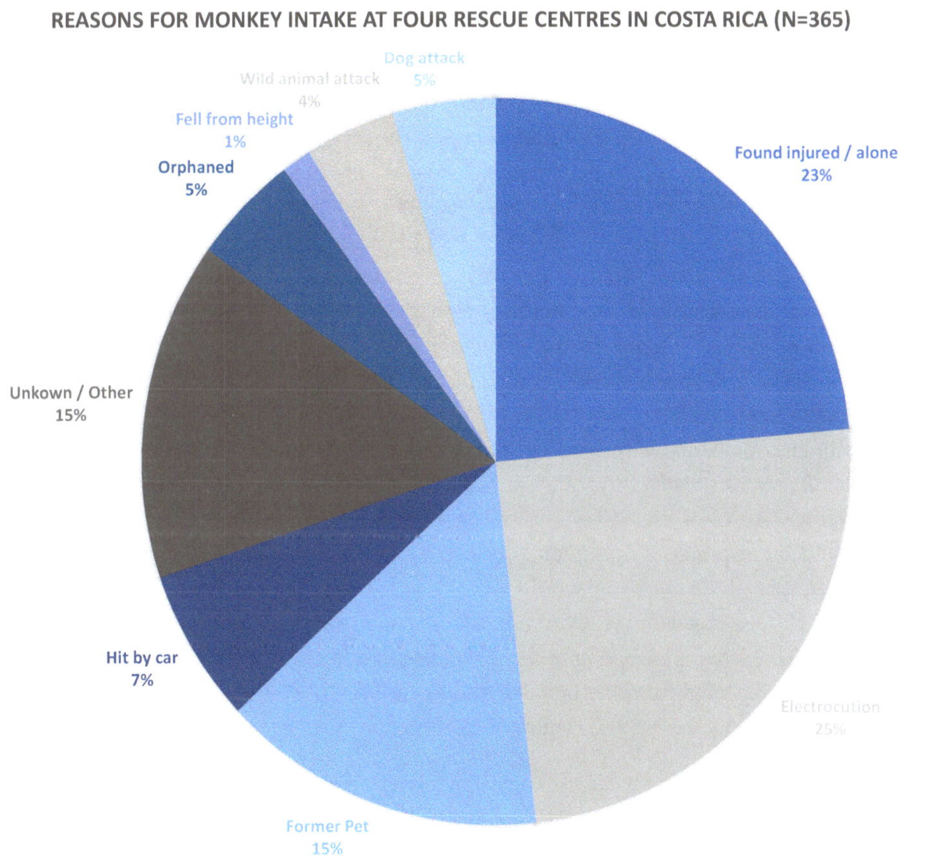

Figure 11.5: Reasons for Monkey Intake at Four Rescue Centers in Costa Rica (n = 365).

in Área de Conservación Guanacaste which is not one of my three focal sites (n = 67). I included data from this additional site in the interest of creating a more robust dataset. I located 365 records of intake for primate individuals to generate this comparative graph.

Animal geographers contend that animals are a social group embedded in complex power relations within "human systems of recognition, ordering, and production" (Hovorka, 2018). Animals are subjects possessing agency, or, the ability to influence relational networks and the environment. In Geiger and Hovorka's exploration of donkey positionality in Botswana, they ask: "how is donkey identity shaped by the ways in which human owners' value, use, and care for them?" (2015, p. 6). Extending this question to the context at-hand, how is monkey subjectivity – their lived experience – shaped by their positionality as either *resident* or *rehabilitant*, and how does this influence the performance of labor? To answer, I bifurcate my discussion of monkey work to grapple with these two positionalities. Whether a monkey is located in the sanctuary or rescue center determines not only their *umwelt*, or inner lifeworld (Dyck, 2012), but also their social networks and agency. Expressions of agency can, as I will come to demonstrate, intertwine with and emerge as forms of labor.

Across the sanctuaries I visited, human-monkey relationships do not appear forced; some individual monkeys display preferences for a particular staff member. Comments such as "this monkey doesn't like men with beards" or "that monkey only likes women and children" are not uncommon. I observed that veterinarians and technicians were met with the most resistance and fear from monkeys, for understandable reasons. When a particular person is disliked by a monkey for no obvious reason, however, it is speculated that the person reminds the monkey physically of someone – likely unkind – from their past. Respecting when a monkey does not like a particular person allows them the agency to choose with whom they interact. Providing opportunities for choice, control, and positive social relationships is an indicator of good welfare (Mellor, 2016). In this way, monkeys are empowered to elect their human "co-workers." This is an example of Coulter's *interspecies solidarity*, since it is founded on a sense of empathy for resident monkeys as victims of trauma – physical or psychological – whose individual preferences should be respected. Consequently, when a monkey chooses to initiate a relational encounter, it is all the more impactful.

To explore monkey subjectivity and agency as it transgresses species boundaries, I invoke *responsible anthropomorphism*, which attends closely "to understandings of non-humans garnered from the practice and experience of co-relationality" (Bear, 2011, p. 302, citing Johnston, 2008). From my field notes:

> At site 2 lives a temperamental male spider monkey notorious for throwing 'tantrums' when a tour guide he does not like enters the sanctuary. His outbursts, screaming and swinging himself against the fence, are so jarring to the peacefulness of the sanctuary that tour groups will not approach his enclosure if he is not in a 'good mood'. My daily observations of the capuchins place me within sight of his enclosure. While I am observing, I see him in my peripheral vision watching me. If I turn towards him, he vocalizes. Feeling emboldened on my last day,

> I took tentative steps towards his enclosure, mimicking his vocalizations. After about ten minutes of 'talking with him', I felt his tail suddenly wrap around my wrist. The action was so unexpected it made me jolt backwards! I looked down to see his 'fifth limb' retracting back through the fence between us. I had severed our physical connection, while simultaneously I sensed the severing of another, unspoken connection. He looked – dare I say – taken aback? And so as not to offend him, I stood there and continued to speak with him. I cannot know his intent or reaction to what passed between us at this interface, but his small gesture for contact was incredibly moving. (7/8/2019)

The phenomenon of monkeys trespassing the artificial boundaries we place them in is not unique to this anecdote; I will explore this idea further.

Emotion work

Emotion work underlies conceivably all other forms of monkey labor. In this section, I consider how the performance of territoriality is indeed labor, though it is an innate suite of behaviors. How is our understanding of territoriality recast when the territory is stationary and bounded by fencing? Before moving into my behavioral research, I present accounts of monkey individuals who are noteworthy for their emotion work, exposing themselves to conflict in the interest of maintaining their unique network of relations.

Monkeys are reputed for escaping captivity, necessitating a series of complicated locks and shift cages to secure them. The male spider monkey I spoke of has earned the nickname "Houdini," since he has escaped a record-breaking 13 times in his 8 years at the sanctuary (Interview, K28). The former sanctuary manager tells me that years ago "Houdini" and his older female companion escaped their enclosure together while she was doing her early morning rounds. In her own words:

> The sanctuary animals had not yet been fed, and I didn't want to leave them all hanging [. . .] plus I thought I might be able to lure [Houdini] with food, and he'd be less nervous with just me around. So, I took all the food down in a big bin & tried to push it around from cage to cage – while [the female spider monkey] had already climbed up onto me and just rode around on my hip! [Houdini], not to be excluded, followed us around, considering the logistics of trying to get on me, as well. But luckily, he accepted simple hand-holding instead, with occasional food thrown away from me to distract him while I tried to open cage doors and slide food in [. . .] eventually [the female] let me walk her back into her enclosure, but [Houdini] needed reinforcements called in for darting! (K26, 07/08/19)

This is work done *with* and *for* animals (Coulter, 2015) – subsistence work done literally hand-in-hand with one monkey while another clings to her hip. In this material and emotional encounter, both monkey and human are "mutually vulnerable" to injury (Parreñas, 2018). Practically speaking, this situation was an impediment to the feeding task. I posit however that the monkeys' recognized the *care* and *body work* the manager was performing, and, rather than escape into the surrounding jungle, they remained with her. One cannot know the monkeys' intentions, but as Coulter notes, animals "subvert[ing] their feelings or desires in order to meet the needs of people [. . .] provides yet another example of animals' emotion work" (2015, p. 76). The level of trust demonstrated in this event by both parties also speaks to *communication* work. The "shared language" shaped in their relational encounters over time had led the monkeys to recognize her as a provider of care (Coulter, 2015).

Despite best efforts to place protective borders around wildlife, animals will find a way to breach them (Main & Chambers, 2014). Consider the life-world of a wild male capuchin at Site 3, who makes almost-daily visits to three resident female capuchins. Nicknamed "Dennis the Menace," he began visiting the sanctuary 2–3 years ago, demanding attention from the captive females in the form of grooming, and has even attempted to copulate with them through the enclosure fencing.

As a result, I began to think of Dennis as part of their troupe. The females share their food with him and take turns grooming him – performing *emotion* and *care work* (Figure 11.6). In return for this affiliative exchange, he provides free-ranging vigilance and protection from wild interlopers. When a wild male disturbed the peace of the sanctuary one quiet morning, Dennis threatened him in coalition with the three captive females; a performance of *body work* (risking his own safety), *care work* (protecting the residents), and *communication work* (territorial coalition with the females) (Figure 11.7). In chasing off the wild male, Dennis labored in the interest of ensuring the wellbeing of his adopted troupe, and in exchange he receives care. Recalling the "mutual vulnerability" of keeper-primate relationships (Parreñas, 2018), I argue the same applies to Dennis' relationship with the residents; he is *performing keeper*.

It is not uncommon for wild capuchins to fight with captive capuchins by reaching through the fencing of the enclosure, and it can lead to serious injury and death. At Site 1, a wild male nearly severed the arm of the alpha male (Figure 11.8), and at Site 2, a wild male injured the alpha female's finger so badly it was left with permanent nerve damage (Figure 11.9). These "war wounds" have serious consequences for the health and welfare of primates, yet are difficult to control since wild monkey raids strategically occur when there are no staff around the sanctuary.

Captivity is permeable. Monkeys are "transgressive boundary creature[s]" (Instone, 2004, p. 37, cited in Main & Chambers, 2014), who move across the enclosures we build around them, "leading to 'out of place' phenomena" (Philo, 1998, p. 52, cited in Main & Chambers, 2014). For residents, whose lives have been micromanaged and sanitized of their wildness – literally and figuratively – the ability to transgress anthropogenic boundaries represent one of their last vestiges of agency. Parreñas claims that "enclosure" is "shorthand for a shared interface of loss between displaced wildlife who have nowhere else to go" (2018, p.28). If this is the case, then monkeys regularly transgress this "interface of loss" by engaging in affective encounters with humans and wild conspecifics through the mesh (Figure 11.10). This is a sign of hope that, while

Figure 11.6 and 11.7: Dennis being groomed through the enclosure by the alpha female.
Dennis performing a coalitionary threat face with a resident female.
Source: Author.

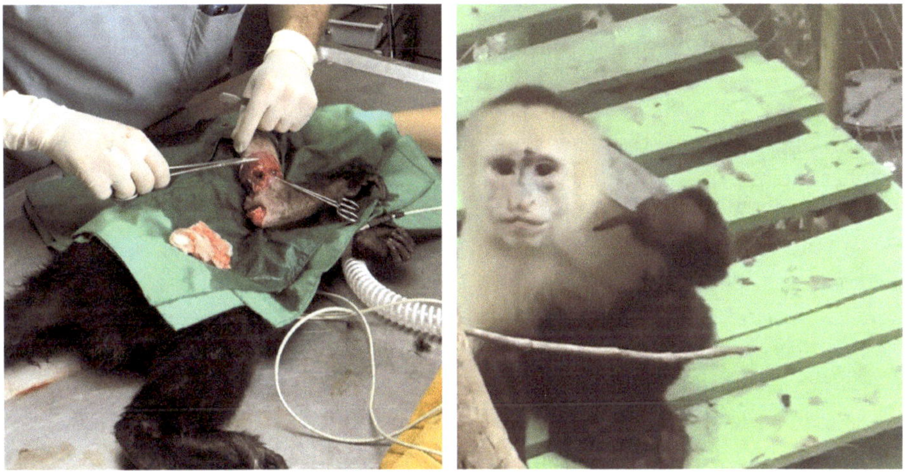

Figure 11.8 and 11.9: The alpha male at Site 1 receiving surgery after a wild monkey attack.
The alpha female at Site 2 with her permanently-extended finger, clutching her mirror which she carries everywhere to look into.
Source: Site 1. Used with permission.

neutered of their ecosocial relations, they will persist in performing works of *care, body, communication,* and *emotion* even when it requires moving beyond that interface of vulnerability signified materially by a fenced enclosure.

The *emotion, body,* and *care work* involved in defending one's troupe while bound in a permeable, panopticon-like enclosure is a reality for these individuals. I conducted 39.6 hours of behavioral observation of resident capuchins across the three sites, observing their behavior before, during and after tourists approached

Figure 11.10: A male capuchin reaching out to ask me for the flower I was holding.
Source: Author.

the enclosure. I chose capuchins for comparison purposes, since they were the only species common across the three sanctuaries.

Site 1 offers 12 tour slots per week, Site 2 offers 6 and Site 3 offers 24. It is important to note, however, that each tour slot is not always filled, depending on the weather and season. I performed focal and group scan observational protocol (Martin & Bateson, 2007) between 9 and 11 AM, during which morning tours were conducted in the sanctuary. Tour groups stopped in front of the capuchin enclosure for about 5 to 15 minutes, depending on the crowd and tour guide, and at least 1 meter away from the fencing. This allowed me to gather observational data before, during, and after tourists approached the enclosure.

During my observational sessions, I noted increased vigilance amongst capuchins in the minutes leading up to tourist approach, when the sound of the tour guide and group becomes increasingly audible. There seemed to me a routine vigilance amongst the capuchins at Site 1, in particular. The alpha male and high-ranking female "L" performed the most vigilance of the group of five, walking the perimeter from the highest branches for ideal vantage points of the incoming tour groups; their movements uncannily resembled that of sentry guards.

Vigilance is a task which fully engages mental and physical faculties, and can lead to more territorial and aggressive behaviors; open-mouthed threat faces, vocalizations, throwing gravel, banging stones and limbs – warning tourists whose territory they are trespassing upon (Figure 11.11).

The occurrence of "tail banging behavior," in which the individual slaps their tail against a hard surface to create a loud sound, occurred at all three sanctuaries. It usually was performed by the alpha, and varied with the number of tourists present and their noisiness as they approached (Figure 11.12). In my field notes, I labeled this behavior "drum," because I overheard a tour guide joke to his group "Look! He is

Figure 11.11: Resident capuchin banging a plaque at a group of tourists and throwing gravel in the second image.
Source: Author.

drumming for you!" This is an example of how *emotion* (vigilance) and *body work* (territoriality) can be interpreted as entertainment by tourists.

During one notable session at Site 1, the alpha male and female "L" banged their tails on the enclosure's roof 14 times during a session of 60 minutes (Figure 11.12). Overall, the most occurrences of tail-banging happened at Site 1, where the enclosure was the smallest of the three sites, held the most individuals, and had nowhere in the enclosure where capuchins could be out of sight.

Site 2 had the most appropriately-enriched enclosure and offered the least tours per week. It is also geographically isolated from the city and located on the edge of a (albeit, fragmented) jungle. Therefore, there is no ambient urban noise. Given this, it follows that the capuchins of Site 2 performed tail-banging the least, and usually appeared unbothered by tourists. Site 3 falls in between Sites 1 in 2 in terms of tail-banging occurrences, and interestingly, lower-ranking female "C" was the most attentive to tourist presence – banging a plaque located on the outside of the enclosure and throwing gravel through the fencing at tourists (Figure 11.12).

Though territorial behavior occurs in the wild, it is recast in the artificiality of captivity. The labor involved in these performances span multiple valences: it is *body work* (physical effort), *care work* (protecting their troupe), and *emotion work* (cannot relax, heightened stress/frustration at times).

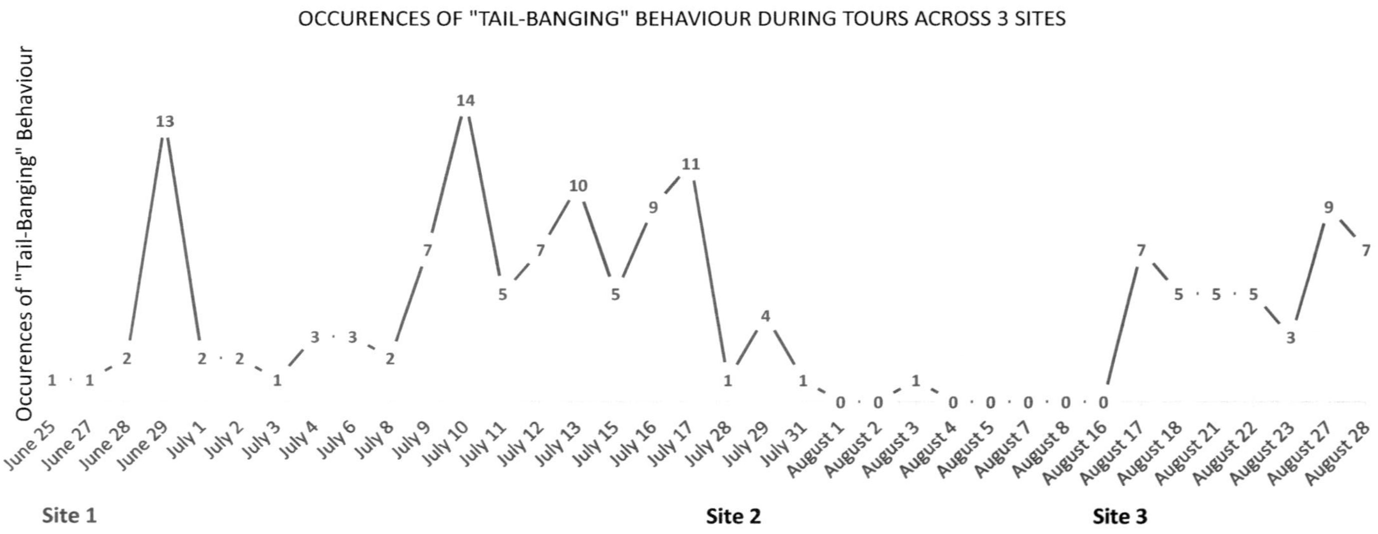

Figure 11.12: Occurrences of "Tail-Banging" Behavior During Tours Across 3 Sanctuaries in Costa Rica.
Note: Sessions lasted 60 minutes at Site 1, and 72 minutes at Sites 2 and 3.

Educational and entertainment work

Having discussed some of the *emotion* and *care* labor resident monkeys perform, I focus now on the enmeshed *entertaining, educational, ecological* dimensions in which they work, specifically as living tourist attractions and spokesanimals. Koko the Gorilla, a famous spokesanimal, became "a kind of ambassador for her species [. . .] participat[ing] in broader educational and conservation efforts [. . .] by offering people a glimpse into gorillas' minds and thus fostering understanding and empathy"; as such, she was "framed (accurately) as an active, multidimensional social agent" (Coulter, 2015, p. 111). Monkeys are inherently entertaining to watch, but what differentiates the *entertainment work* of a circus monkey versus a sanctuary monkey is that in the latter, monkeys are allowed to be "multidimensional social agents;" their instinctive behaviors are not suppressed, but rather encouraged. A tour guide I spoke to had this to say:

> I tell people on my tour that [the monkeys] are my coworkers. That's because I respect them. They're not *my* monkeys. [. . .] When I talk about monkeys [in front of the enclosure] and they are behind me, I think they come next to me and pose for the picture. (K14, Interview)

The perception of monkeys as intentionally "posing for pictures" or making noise for tourists recurred throughout various interviews. Their role as entertainers of their own accord reframes them as "active subjects with their own agency as opposed to static objects without agency" (Main & Chambers, 2014, p. 60). The potential stress or disruption caused by a constant flow of tourists is an acknowledged trade off at sanctuaries. For better or for worse, "performing monkey" is literally their meal ticket. Furthermore, it is not only tourists who find monkeys entertaining. I conducted photo-interpretive interviews with twenty-five Costa Ricans living around my study sites. When shown images of monkeys climbing peoples' shoulders and receiving food – two interactions which, while illegal, are known to still occur – most responses were "gracioso" (funny), "divertido" (fun), and "amigable" (friendly).

When an attraction incorporates monkeys' natural charisma with the storying of their individual lives, it increases the chance that tourists will experience affective encounters and take an interest in their conservation. This is the case on sanctuary tours, which have strong, sobering environmental overtones. During a tour of Site 1, a guide related the seriousness of the illegal pet trade to our tour group, saying:

> People want baby monkeys as pets. So, you must shoot the mother; that is how you get the baby. But the mother, maybe she still could have had five more babies. So, when you kill the mother, you are not just killing one individual, but you are killing the five babies she could have had as well. (K15, Interview)

In this quote, the guide makes evident (1) the psychological trauma monkeys experience prior to rescue, (2) how anthropogenic pressures have multigenerational impacts on their populations, and (3) the "ecosocial reproduction" of wild monkeys (Coulter, 2015, p. 64). The ability to perform ecological labor – in which wild individuals live and reproduce within a relational web that maintains the biodiversity and resiliency of the ecosystem – is essentially halted when monkeys are removed and relocated to rescue centers. The work of care which staff perform to rehabilitate these individuals is, in essence, to restore the "wild" agency of the monkey such that they can perform ecological labor in the wild. If they cannot demonstrate this ability, or have come from a captive past, they are not released and are transferred to the sanctuary.

Ecological work

The relationship between staff, volunteers, and tourists vary between *resident* and *rehabilitant* monkey positionalities. This fundamentally shapes their lives, labors, and their interspecies relationships with humans. Hands-on relationships with rehabilitants, when they are no longer infants, are discouraged since they may lead to habituation. From a monkey's perspective, rehabilitators may be "mother" for a short time when they are dependent on humans for care work as infants or intensive care patients. Even then, the monkey is given a stuffed animal as a hug-substitute for the cling-reflex while receiving hands-on attention. The work of maternal care in this case is materially mediated to discourage mutual, interspecies attachments.

As rehabilitant monkeys prepare for release, they can no longer recognize humans as potential benefactors – this is a key marker of being "ready to be (re)wild(ed)." Rehabilitators transition from "mother" to "stranger." To achieve this, "misanthropic practices" may be employed, as defined by Collard: "actions and routines designed to instill in animals fear and even hatred of humans, driven by a pronounced lack of faith in human beings or, more specifically, by the idea that contact with human beings puts animals at high risk of death, suffering, and recommodification" (Collard, 2014, p. 154). The practices she describes include spraying water into enclosures, keeping one's distance and discouraging interspecies touch – all of which I observed at my field sites. I argue that the severing of human-monkey connections formed in the rehabilitation process requires suppressing the desire for physical and emotional contact with one another. This is surely a form of emotional labor, since following Coulter, it requires monkeys to "behave in specific ways, and to subvert their feelings or desires in order to meet the needs of people" (2015, p.76) – our need being, in this instance for them to be "rewilded." The bisection of the human-monkey relationship, however, is not always a clean break.

In addition to the three sites I focus on, I visited two rescue centers that specialize in orphaned monkeys. Both centers arose similarly; decades ago, two expat women began rehabilitating monkeys out of their respective homes, and now run rescue operations. Coulter observes that "communication work, emotional work, body work, and sometimes even dirty work intersect as part of care work (as is true for their human coworkers)" (2015, p. 75). Through my interviews with the two owners, the maternal work of care and all its intersections with other labors, was centered. Curiously, both owners had similar experiences of releasing rehabilitants who return to visit them. For example, a released female came regularly for two years until she had a baby. She visited her "monkey mama" (the owner) one last time, making affectionate noises and "showing off her baby" before departing (K1, interview). At the other center, I witnessed an encounter which I describe in my field notes:

> The tour is over and the other tourists have left. I sit down with the owner for an interview. Her enthusiasm is infectious. It is clear that rehabilitating monkeys is a labor of love; she uses maternal instincts to rehabilitate and release monkeys. Some of the earliest releases took place around the center. As we speak, a wild male howler monkey moves across a high branch overhead. He stops above my interlocutor and scratches himself, before swinging down onto her bench. She doesn't acknowledge him, even as he curls up behind her back so that I can't see him. He wraps his long, elegant tail around her waist. She pays him no attention, and tells me that despite being released, he will still visit from time to time. (K7, interview)

The work of monkey rehabilitation is an understudied, highly divisive, and fraught with fragility. A common thread which emerged from my interviews is that this work is undermined with professional jealousy. It is hard for those at the heart of it to trust outsiders, such as myself. To witness this once-orphaned howler quietly embrace whom he perceives to be his mother-figure was intensely emotional. Yet another moment of shared vulnerability; the owner trusting that I will not judge this interaction, and the howler trusting that we will not harm him. This is an affiliative space of cross-species comfort and connection. In this moment, I was reminded of Parreñas' (2018) call to address the harshness and violence of some rehabilitation work, asking, "can we instead embrace the vulnerability of sharing our lives together, however fleeting those moments might be?" (p. 3).

Maternal themes were evident across many interviews I conducted. One veterinarian told me that releasing an orphaned monkey, whom she had raised from infancy, moved her to tears: "I felt the joy of having a baby and we released her" (K36, Interview). A wildlife nursery manager told me that much of her work feels instinctual: "I feel like I'm doing the right thing with my life [. . .] It's one of the most draining parts of working with [orphans], thinking about what they need, the suffering they're going through . . . " (K26, Interview). In these anecdotes, we see the merging of emotional and ecological labor performed by both rehabilitators and rehabilitants.

Released monkeys signify hope and optimism; their wild agency has been restored, and they no longer rely on rehabilitators for subsistence. They are subsumed back into the local ecology in which they perform crucial ecosystem services (Estrada et al., 2017) as seed dispersers. Their release means the re-initiation of their role as ecological laborers – a decommodifying transformation. In her case study of a rescue center in Guatemala, Collard writes that rescue centers have a "two-fold" duty: to rebuild their "wild lives" and "to undo their commodity lives" by "divest[ing] them of their dependence of human inputs and of their pet-like behaviours" (2014, p. 152). Collard remarks notably that such work "put[s] animals back together and take[s] commodities apart" (p. 152). Following Collard (2014), however, this is not necessarily a process of "renaturing" or rewilding as it is "another production of nature" (p. 155). Furthermore, it may also be a decolonial process, a "fundamental reorientation toward other, especially non-human others, in which we accept the risk of living together" (Parreñas, 2018, p. 27).

Can *resident* monkeys perform ecological labor? I posit yes, because their positionality as educational ambassadors and attractions not only funds "subsistence work for themselves" but "for/with others" (Parreñas, 2018, p. 62), in particular the conservation of their wild conspecifics. Unable to be released, or "decommodified," residents live their lives subject to multiple gazes: the tourist's gaze, keepers' gaze, and social media's gaze (Figure 11.13). The latter is of particular interest because at all sanctuaries and rescue centers I visited, photos of their animal residents, accompanied by anecdotes or stories, are regularly uploaded onto social media. In this way, their animal bodies are transformed into *virtual bodies* (Whatmore & Thorne,

Figure 11.13: The author observing sanctuary capuchins while a tour is taking place.
Note: The capuchin individual appears in a sort-of panopticon, visible from all sides.
Source: Rubén Vargas. Used with permission.

2000) based on their *nonhuman charisma* (Lorimer, 2007) and potential to attract tourists, donors, or conservation-minded "followers."

In Main and Chambers' (2014) exploration of "Sirocco," an Australian kakapo and conservation icon, they comment that

> Sirocco's fleshy presence is distributed through a new topology of encounter where his transgressive nature and un-parrot like fascination with humans is actively celebrated and used as a means of further endearing Kakapo to the general public. (p. 71)

This observation applies to the context-at-hand, with Facebook and Instagram as the "new topology of encounter" where sanctuaries generate a virtual following by posting updates on residents and rehabilitants, sharing requests for funding or donations, and advertising job and volunteer opportunities.

While monkeys are aware of the tourist gaze from their experience with daily tours, they unknowingly labor in virtual spaces as ambassadors for their species. Given their environmental education mandate, sanctuaries have essentially transformed monkey bodies into material and virtual spokesmonkeys for anthropogenic extinction; living "hazard" symbols for the "arrested autonomy" (Parreñas, 2018, p. 23) of those whose wild lives have been interrupted by humans.

Conclusion

In a way, sanctuaries and rescue centers are a mirror on ourselves in that it forces us to grapple with the reality of how our actions (and inactions) have gravely influenced the lives and agencies of wild non-humans. It is a site in which we come face-to-face with morbidity, where the threat of extinction hangs overhead. These are places of mourning: a hospice (Parreñas, 2018) for rehabilitant monkeys who regularly arrive on death row; where resident monkeys, scrubbed of their "monkeyness," exist in the space between wild and tamed.

These are also places of hope. The labor of wildlife rehabilitation opposes colonial legacies, returning animals back to their natural ranges, and caring for those in captivity who cannot return to the wild. Through the interspecies labor *with* and *for* these monkey individuals, expressions of agency emerge. Monkeys elect to have affective relationships with human staff, when this requires *emotion* and *communication work* from both parties. Further, monkey *body* and *care work* uphold social networks and safeguards the wellbeing of humans and conspecifics alike. As *educational* ambassadors and *entertaining* living and virtual figures, both residents and rehabilitants generate critical funding for their own husbandry and conservation. Raising funding and public awareness of the conservation issues plaguing their wild counterparts, promotes the ability of wild populations to perform *ecological* labor in the form of ecosystem services. Monkeys are sentient beings who participate in forms of labor which, though distinct, bleed into each other – all underlaid by the emotional work of care that occurs at the human-monkey interface of rehabilitation.

References

Alexander, S. E. (2000). Resident attitudes towards conservation and black howler monkeys in Belize: The Community Baboon Sanctuary. *Environmental Conservation, 27*(4), 341–350.

Bear, C. (2011). Being Angelica? Exploring individual animal geographies: Being Angelica? *Area*, 43(3), 297–304.

Birke, L., Hosey, G., & Melfi, V. (2019). "You Can't Really Hug a Tiger": Zookeepers and Their Bonds with Animals. *Anthrozoös, 32*(5), 597–612

Brockington, D. (2002) *Fortress Conservation: The Preservation of the Mkomazi Game Reserve, Tanzania*. Bloomington, IN: Indiana University Press.

Carder, G., Plese, T., Machado, F., Paterson, S., Matthews, N., McAnea, L., & D'Cruze, N. (2018). The Impact of 'Selfie' Tourism on the Behaviour and Welfare of Brown-Throated Three-Toed Sloths. *Animals, 8*(11), 216.

Cohen, E. (2013). "Buddhist Compassion" and "Animal Abuse" in Thailand's Tiger Temple. *Society & Animals, 21*(3), 266–283.

Collard, R.-C. (2014). Putting animals back together, taking commodities apart. *Annals of the Association of American Geographers, 104*(1), 151–165.

Compton, N. B. (2019, July 12). How to find a real animal sanctuary. *Washington Post*. Retrieved from https://www.washingtonpost.com/travel/tips/how-find-real-animal-sanctuary/

Coulter, K. (2015). *Animals, Work, and the Promise of Interspecies Solidarity*. New York: Palgrave Macmillan US: Imprint: Palgrave Macmillan.

Das, M. & Chatterjee, B. (2015). Ecotourism and Empowerment: A Case Analysis of Bhitarkanika Wildlife Sanctuary, Odisha, India. *IIM Kozhikode Society & Management Review, 4*(2), 136–145.

D'Cruze, N., Machado, F. C., Matthews, N., Balaskas, M., Carder, G., Richardson, V., & Vieto, R. (2017). A review of wildlife ecotourism in Manaus, Brazil. *Nature Conservation, 22*, 1–16.

Dore, K. M., Riley, E.P, & Fuentes, A. (2017). *Ethnoprimatology: A Practical Guide to Research at the Human-Nonhuman Primate Interface*. Cambridge Studies in Biological and Evolutionary Anthropology. Cambridge and New York: Cambridge University Press.

Dore, K. M. (2018). Ethnoprimatology without Conservation: The Political Ecology of Farmer–Green Monkey (Chlorocebus sabaeus) Relations in St. Kitts, West Indies. *International Journal of Primatology, 39*(5), 918–944.

Duffy, R. (2008). Neoliberalising Nature: Global Networks and Ecotourism Development in Madagasgar. *Journal of Sustainable Tourism, 16*(3), 327. https://doi.org/10.2167/jost748.0

Dyck, H. V. (2012). Changing organisms in rapidly changing anthropogenic landscapes: The significance of the 'Umwelt'-concept and functional habitat for animal conservation. *Evolutionary Applications, 5*(2), 144–153. https://doi.org/10.1111/j.1752-4571.2011.00230.x

Estrada, A., Garber, P. A., Rylands, A. B., Roos, C., Fernandez-Duque, E., Di Fiore, A., . . . Li, B. (2017). Impending extinction crisis of the world's primates: Why primates matter. *Science Advances*, 3(1).

Fennell, D. A. (2012). *Tourism and Animal Ethics*. London: Routledge.

Fragaszy, D. M., Visalberghi, E., & Fedigan, L. M. (2004). *The complete capuchin: The biology of the genus Cebus*. Cambridge: Cambridge University Press.

Fuentes A. & Wolfe, L. D. (2002). *Primates face to face: the conservation implications of human-nonhuman primate interconnections*. Cambridge: Cambridge University Press.
Fuentes, A. (2012). Ethnoprimatology and the Anthropology of the Human-Primate Interface. *Annual Review of Anthropology*, 41(1), 101–117. https://doi.org/10.1146/annurev-anthro-092611-145808
Geiger, M. & Hovorka, A. J. (2015). Animal performativity: Exploring the lives of donkeys in Botswana. *Environment and Planning D: Society and Space*, 33(6), 1098–1117.
Horton, L. R. (2009). Buying Up Nature: Economic and Social Impacts of Costa Rica's Ecotourism Boom. *Latin American Perspectives*, 36(3), 93–107. https://doi.org/10.1177/0094582X09334299
Hovorka, A. J. (2017). Animal geographies I: Globalizing and decolonizing. *Progress in Human Geography*, 41(3), 382–394. https://doi.org/10.1177/0309132516646291
Hovorka, A. J. (2018a). Animal geographies II: Hybridizing. *Progress in Human Geography*, *42*(3), 453–462. https://doi.org/10.1177/0309132517699924
Hovorka, A. J. (2018b). Animal geographies III: Species relations of power. *Progress in Human Geography*, 030913251877583. https://doi.org/10.1177/0309132518775837
Humane Jobs. (n.d.). Humane Jobs website: https://humanejobs.org/. Retrieved October 31, 2019.
Instone, L. (2004). Situating nature: On doing cultural geographies of Australian nature. *Australian Geographer*, 35, 131–140.
Johnston, C., (2008). Beyond the clearing: towards a dwelt animal geography. *Progress in Human Geography*. 32, 633–49.
Langholz, J. A. & Lassoie, J. P. (2001). Perils and Promise of Privately Owned Protected Areas. BioScience, 51(12), 1079–1085.
Lenzi, C.; Speiran, S., & Grasso, C. (2020). "Let Me Take a Selfie": Reviewing the Implications of Social Media for Public Perceptions of Wild Animals. *Society & Animals*.
Locke, P. (2013). Explorations in Ethnoelephantology: Social, Historical, and Ecological Intersections between Asian Elephants and Humans. *Environment and Society*, *4*(1).
Locke, P. (2017). Elephants as persons, affective apprenticeship, and fieldwork with nonhuman informants in Nepal. *HAU: Journal of Ethnographic Theory*, *7*(1), 353–376.
Lorimer, J. (2007). Nonhuman Charisma. *Environment and Planning D: Society and Space*, *25*(5), 911–932.
Lorimer, J. (2012). Multinatural geographies for the Anthropocene. *Progress in Human Geography*, *36*(5), 593–612.
Main, M. E. & Chambers, C. N. L. (2014). Between "Wild" and "Tame": Placing Encounters with Sirocco the Kakapo Parrot in Aotearoa/New Zealand. *Society & Animals*, *22*(1), 57–79. https://doi.org/10.1163/15685306-12341319.
Martin, P. R., & Bateson, P. P. G. (2007). *Measuring Behaviour: an Introductory Guide*. 3rd ed., Cambridge University Press
McKinney, T. (2010). *Social and Ecological Impact of Anthropogenic Disturbance on the Sympatric White-faced Capuchin (Cebus capucinus) and Mantled Howler Monkey (Alouatta palliata)* (The Ohio State University). Retrieved from https://etd.ohiolink.edu/pg_10?0::NO:10:P10_ACCESSION_NUM:osu1267816356
Mellor, D. (2016). Moving beyond the "Five Freedoms" by Updating the "Five Provisions" and Introducing Aligned "Animal Welfare Aims." *Animals*, *6*(12), 59. https://doi.org/10.3390/ani6100059
Moorhouse, T. P., Dahlsjö, C. A. L., Baker, S. E., D'Cruze, N. C., & Macdonald, D. W. (2015). The Customer Isn't Always Right – Conservation and Animal Welfare Implications of the Increasing Demand for Wildlife Tourism. PLOS ONE, 10(10), e0138939.
Moorhouse, T. P., D'Cruze, N. C., & Macdonald, D. W. (2017). The effect of priming, nationality and greenwashing on preferences for wildlife tourist attractions. *Global Ecology and Conservation*, *12*, 188–203. https://doi.org/10.1016/j.gecco.2017.11.007

Paquet, P. & Darimont, C. (2010). Wildlife conservation and animal welfare: Two sides of the same coin? *Animal Welfare*, 14.

Parreñas Juno Salazar. (2018). *Decolonizing extinction: the work of care in orangutan rehabilitation*. Durham: Duke University Press.

Philo, C. (1998). Animals, geography and the city: Notes on inclusions and exclusions. In J. Wolch & J. Emel (Eds.), *Animal geographies: Place, politics, and identity in the nature-culture borderlands* (pp. 51–72). London, United Kingdom: Verso.

Ramp, D. & Bekoff, M. (2015). Compassion as a practical and evolved ethic for conservation. *BioScience*, *65*(3), 323–327.

Risan, L. C. (2005). The Boundary of Animality. *Environment and Planning D: Society and Space*, *23*(5), 787–793. https://doi.org/10.1068/d359t

Sebele, L. S. (2010). Community-based tourism ventures, benefits and challenges: Khama Rhino Sanctuary Trust, Central District, Botswana. *Tourism Management*, *31*(1), 136–146.

SINAC. "Sistema Nacional De Áreas De Conservación (SINAC)." *¿Sabía Que Costa Rica Cuenta Con 11 . . . – Sistema Nacional De Áreas De Conservación (SINAC)*, 25 Sept. 2018, www.facebook.com/cr.sinac/photos/a.414883101929887/1874992345918948/?type=3&theater. Accessed 8 July 2020

Wallace, G. & Pierce, S. (1996), An evaluation of ecotourism in Amazonas, Brazil. *Annals of Tourism Research*, 23, 4, 843–873.

Whatmore, S. & Thorne, L. (2000). Elephants on the Move: Spatial Formations of Wildlife Exchange. *Environment and Planning D: Society and Space*, 18(2), 185–203.

Wolch, J. & Emel, J. (1995). "Guest Editorial on 'Bringing the Animals Back In'." *Environment and Planning D: Society and Space*, 13: 632–636.

Jill Bueddefeld, Mary Benbow

12 The greening of polar bears: Lively commodities in a climate change economy

Introduction

While the work that animals do in the tourism industry has been under-researched, this chapter contributes to this body of knowledge by exploring the hidden and symbolic work polar bears perform in the context of tourism (Carr, 2016; Duffy, 2014; Fennell, 2011; Kline, 2018; Rutherford, 2011; Young & Carr, 2018). Specifically, this chapter discusses the work that polar bears are made to do as ambassadors, both for their own species and as a symbol for climate change. In order to understand their ambassadorial work, we begin by providing context for polar bear tourism in Manitoba, Canada, and the viewing opportunities at both in situ (wild) and ex situ (captive) sites in this region. These case studies allow for a critical analysis of polar bear "work" and a comparison of the ways in which these different settings can alter our perceptions of the bears. In particular, how humans regard them as stand-ins for complex problems is analyzed. Unlike previous research, which explores the literal work that animals perform in tourism, such as working elephants (Duffy, 2014), camels, cows, or horses which often carry people or their luggage (Duffy & Moore, 2010) or the mobility and safety work that service animals do (Rickly, 2018), this research examines the labor that polar bears perform as ambassadors. While this might less obviously be categorized as work, it is, nevertheless, an essential component of their perceived value as tourism attractions and symbols of broader social discourses around ecological crises.

In recent decades polar bears have become a globally recognized symbol for climate change (Born, 2019) (Figure 12.1). As sea ice decreases in area, forms later in the autumn, and recedes earlier in the spring in the northern latitudes, polar bears are rapidly losing their habitat and their ability to hunt their primary food source: seals. This direct connection between a large charismatic species and climate change makes polar bears ideal candidates for acting as ambassadors for all species affected by climate change. Stories and images of polar bears are used to evoke emotion and to prompt people to change behaviors in order to positively affect climate change. In the context of tourism, however, live polar bears can be witnessed in both wild (in situ) and captive (ex situ) settings, which adds an embodied, lived experience of the species to their previous symbolic encounters. Indeed, many ex situ sites, such as zoos, exist under the premise that polar bears are a part of educational work aimed to teach the public about their wild counterparts and the challenges they face, such as the effects of climate change.

https://doi.org/10.1515/9783110664058-012

Figure 12.1: Oxfam and the symbolic use of polar bears.
Source: WikiMedia Commons.

Against this backdrop, this chapter draws out some of the contradictory practices of polar bear tourism, such as employing polar bears as ambassadors for conservation and climate change while simultaneously encouraging high-carbon emissions travel to ecologically sensitive destinations, thereby directly contributing to climate change and subsequent conservation challenges. Informed by Duffy's work (2000; 2014; 2015), we discuss the problematic ways in which animals are used to do the work of marketing tourism and capitalizing on environmental issues. In particular, we note that last chance tourism is marketed as and capitalizes on "witness[ing] (and consum[ing]) the demise of ecosystems, the extinction of species, or even ecocide, at first hand" (Duffy 2015, p. 533). Moreover, the carbon emissions from long distance travel contributes to climate change to the extent that "the long term sustainability of the polar bear viewing industry under projected climate changes is precarious at best" (Dawson et al., 2010, p. 321). In examining these and other issues, this research uses a case study approach to illustrate human-polar bear relations at multiple scales and settings. Duffy and Moore (2010) argue that contrasting case studies research widens the lens of understanding in comparison to a single case study, by allowing us "to examine the contours, boundaries, challenges and limitations placed on neoliberalism by its encounter with 'nature'" (p. 743). Thus, by comparing both captive and wild polar bear tourism contexts, we delve deeper into complex ethical discourses and render visible the moral geographies that underlie the symbolic role into which the human species has directly and indirectly cast the polar bear as an ambassador species.

Finally, this chapter also discusses the ethical implications of a green governmentality that can be used to understand the ambassadorial work that polar bears do in wildlife tourism. Rutherford (2011) explains that practices of green governmentality "provide us with a lexicon through which to apprehend the world, and they tell us what nature is, how it is under threat, and the appropriate solutions to its crisis" (p. 275). How these practices are created and narrated has important implications for understanding the ways in which the work of polar bears is leveraged within wildlife tourism. It has been observed that tourism is "shifting the moral geographies of wildlife from the utopian confines of the sanctuary or the ark of wilderness" (Whatmore, 2002, p. 31), to animals that are used as "symbols and consumables" in the marketing of destinations (Huijbens & Einarsson, 2018, p. 23).

Wildlife tourism: The work of captive and wild animals

This section discusses a short history of in situ and ex situ wildlife tourism experiences in order to provide an understanding of how modern zoos and wildlife tourism experiences have evolved, traditionally using animals to do the work of entertaining and educating the public, thereby rendering nature legible to the public by telling particular stories about animals (Carr, 2016). Taking a critical theory approach, the ramifications of creating places and experiences where animals act as ambassadors to educate people about environmental issues will be explored (Born, 2019; Derrida & Wills, 2002; Rutherford, 2011).

Seeing wildlife and ordering nature

We begin with an examination of the evolution of captive animal tourism, as some of the earliest forms of wildlife tourism took place ex situ, where wildlife was captured, tamed, or killed and stuffed in order to be made visible for human interest (Fennell, 2012). The evolution of zoos reflects changes in societies and cultures and how they have constructed nature. Ginsberg (1993) describes changes from the 19th century menagerie, with their taxonomic theme, to the ecologically themed living museums of the 20th century, culminating with conservation centers of the 21st century with environmental themes (Uddin, 2015). These themes address the different concerns of each era beginning with earlier zoos' emphasis on the clear display of animals, then the 20th century concern of research, public education, and captive propagation, to contemporary concerns of biodiversity, conservation, and species survival with the continued emphasis on public education (Ginsberg, 1993; Uddin, 2015). Zoos often use this perceived progress as a defense against criticisms, including the debates on the ethics of keeping animals in captivity, whether modern zoos can sufficiently provide for all of an animal's welfare, and the degree to which education in zoos is effective (Mason, 2000). That zoos have changed over time is an insufficient defense in and of itself; all public cultural institutions, such as museums, libraries, or galleries have changed over time (Uddin, 2015). In addition, changes in zoos have not occurred at the same rate in all parts of the world, and there are still zoos that are collections of animals kept more for entertainment than for any other reason, these often being referred to as menageries, petting zoos, or even roadside attractions. Disconcertingly, some unaccredited attractions operate under misleading names such as "safaris" or even "zoos" and offer dangerous wildlife petting opportunities and perpetuate captive breeding of wild animals.

It has been argued that modern zoos that meet accreditation standards can make contributions that distinguish them significantly from menageries and zoos of the past (Roe, McConney, & Mansfield, 2014). Using an array of architectural, technological, communications, interpretive, and planning tools, accredited modern zoos consciously construct nature and mediate human-animal relations (Hallman & Benbow, 2007; Roe et al. 2014). Architectural and construction advances allow complex exhibits that can provide, for example, climate-controlled environments, water filtration systems, safe animal handling facilities, and innovative and attractive exhibits (Hallman & Benbow, 2007; Mason, 2000). Technological innovations and computer technologies can link to remote locations through web cameras, extending medical technologies to diagnose and treat animal disease and monitor genetic characteristics. Communications and interpretive programming within zoos are mediated to offer both synchronous and asynchronous engagements and create messages to visitors, which support institutional goals such as education, environmental awareness, and other forms of information (Moss, Esson, & Bazley, 2010; Roe et al., 2014; Stoinski et al., 2002). Lastly, modern zoo exhibits undergo extensive periods of planning to ensure that exhibits, and by virtue of that, the animals themselves, convey messaging to also suit institutional goals. Indeed, many accredited zoos today exist in part due to in situ conservation mandates (Moss et al., 2010; Packer & Ballantyne, 2016; Roe et al., 2014; Stoinski et al., 2002; Tribe & Booth, 2003).

In contrast, wildlife tourism in situ is a relatively contemporary phenomenon. Early forms of wildlife tourism consisted largely of colonial hunting trips, where wealthy Europeans would seek exotic game and consume wildlife as trophies to bring back as evidence of their power and dominion (Rutherford, 2011). Here, consumptive patterns of wildlife tourism demonstrated the effects of the tourist gaze in perpetuating power dynamics, which created a clear hierarchy of human and non-human life (Yudina & Grimwood, 2016). With the rise of the industrial revolution, the middle class, and tourism, there was a corresponding increase in the complexity of wildlife tourism. Ironically, in situ tourism where visitors flock to see wildlife did not fully develop, at least in North America, until delineated "wild spaces" were created with the establishment of national parks (Cronon, 1996). Cronon (1996) describes both a romanticized concept of wilderness where "adventurers," men from elite classes of society who had not had the difficulties of working land themselves, considered the wilderness to be the "ultimate landscape of authenticity" (p. 16), and a dualistic version of wilderness where people were outside or separate from nature.

This dualistic perspective is at the heart of creating a problematic divide between nature and people: "Wilderness dualism tends to cast any *ab-use*, and thereby denies us a middle ground in which responsible use and non-use might attain some kind of balanced, sustainable relationship" (Cronon, 1996, p. 21). This results in the designation of wild and exotic places, in the form of parks, that are close to home, but still far enough away to be distinguished, in the public's view, from everyday life (Rutherford, 2011). In other words, once humanity had been successful enough at removing wildlife from the landscape, designated wild places had to be created in order for tourists to be able to go and see wildlife (Cronon, 1996; Rutherford, 2011). In this way, the creation of wildlife tourism can be seen to be in direct proportion to the moral ordering of human and nature dichotomies and represents a visible shift in the animal gaze (Rutherford, 2011). Power dynamics have evolved through this moral ordering, which the human species ascribes to other non-human life forms, as is evidenced by the shift in wildlife tourism from literal forms of consumption represented in exotic animal or big game hunting, to ocular forms of consumption captured instead by the camera lens (Kline, 2018; Lemelin, 2006; Rutherford, 2011). Lemelin (2006) refers to the ocular consumption of polar bears as part of the tourist gaze, and certainly this transition is part of a literal gaze – where the concept of the gaze is instrumental in laying open the power relations of the systems, which create particular ways of seeing.

The wildlife gaze

The concept of the gaze draws upon the work of Michel Foucault who described how the power and discourse of doctors was, in fact, an act of seeing which was justified and supported by medical institutions and structures (Foucault, 2008; Urry & Larsen, 2011). The gaze describes not only how seeing is informed and normalizes a particular discourse, but also how the impact of being gazed upon further influences subsequent power dynamics (Foucault, 2008). The tourist gaze illustrates the ways tourists' perceptions and attitudes, and the institutions these represent, create and affect that which is gazed upon (Urry & Larsen, 2011). The tourist gaze has power directly through the places and experiences chosen, and indirectly through the ways these choices shape the tourism industry (Urry & Larsen 2011).

In wildlife tourism, both in situ and ex situ, tourists may bring an array of preconceived ideas and attitudes that mediate their gaze (Rutherford, 2011). In both cases, the tourist gaze is carefully framed within the experience, where impressions of wildness are co-mingled with confinement to ensure the safety of the visitors. For example, tourists visiting Churchill assume their experience of polar bears in the wild (or in situ) is more authentic than seeing polar bears in the controlled environment of the zoo (Bueddefeld, 2019). However, polar bear viewing in Churchill it is not without human influence, as tourism operators intentionally take the tourists to locations of polar bear sightings. Specifically, this wildlife gaze within tourism creates a sense of what is and what is not wild.

Braverman (2012) considers that in zoos "the act of seeing animals undermines their wildness" (p. 75), because an animal in the wild would avoid being seen by humans but has no choice in a zoo. Zoos intentionally create exhibits to construct verisimilar nature, with the visitor's gaze complicit in that construction (Rutherford, 2011). Comparing the gaze at the zoo with that in Churchill therefore requires a compromising of authenticity, wildness, and perspectives. "While these gazes may virtually be harmless or non-consumptive, they are predicated on ontological ideas (i.e. the control of nature) and dependent on infrastructures (airports, rail, highways, accommodations, sewers, water systems), which may or may not be sustainable" (Lemelin & Wiersma, 2007, p. 39). Zoos are "not a precise simulacrum of wild nature" (Braverman, 2011, p. 819) because their design must include signs, safety features, and elements to control animal movements and behavior. In addition to the general goals of modern zoos (education, research, conservation, and recreation), zoos design exhibits with specific supporting objectives that reinforce these goals by connecting their agendas to specific social and ecological issues, such as climate change (Moss et al., 2010; Packer & Ballantyne, 2016; Roe et al., 2014; Stoinski et al., 2002; Tribe & Booth, 2003). The wildlife gaze in these spaces can then be understood as directing both physical gaze and movement, as well as the social movements bound up in the conservation messages and consumerism inherent in the visitor experience (Rutherford, 2011).

The historical analysis of zoos has "tended to focus on individual institutions and to emphasize the power relations implicit in the human gaze at caged animals, interpreting it as symbolic of imperial power over colonial subjects" (Hanson, 2004, p. 8). While that privileging of colonialism is not an overt goal of wildlife tourism today, the remnant of that power remains (Uddin, 2015). Both zoos and wildlife tourism settings increasingly attempt to bring diverse constructions of nature from different cultures, while dealing with the challenge of avoiding cultural appropriation. In addition, the stories of individual animals are often used to encourage visitors to make connections to the animals while the origins of other animals may be used to convey a valued narrative (Packer & Ballantyne, 2012). As associated with tourist destinations, animals function as geographic markers that feature in the travel narrative, becoming an iconic symbol of each place (Huijbens & Einarsson, 2018). Animals that, in their wild (in situ) locale, are experiencing environmental threats such as habitat loss, climate change, or disease, are used to convey that narrative (Packer & Ballantyne, 2012). In this respect, it is intended that the attraction of visitors to the animal and exhibit experience can mediate and consolidate an understanding of the issues (Mason, 2000).

While wildlife tourism has strong links to the growing interest in the ecological impacts of tourism, care must be taken to ensure that this does not merely become a green form of commodification (Collard, 2013; 2014; Duffy & Moore, 2010; Markwell, 2015; Rutherford, 2011). Having good intentions for conservation efforts does not necessarily mean that ethical implications have been fully considered. More specifically, as this chapter examines, the use of the polar bear as an ambassador for climate change challenges tourists to continue to encounter the animal as they make various decisions with regard to climate change (Born, 2019; Manzo, 2010). The power dynamics of intentionally centering polar bears as ambassadors in the climate change narrative reveal that much more is at stake for them as lively commodities in the greening of the tourism economy.

Polar bear tourism in Manitoba

The Canadian prairie province of Manitoba is world-renowned for polar bear tourism. On the coast of the Hudson Bay, nearly 1000 kilometers north of the provincial capital Winnipeg, lies the town of Churchill. Churchill promotes itself as "the polar bear capital of the world" and has a reputation as one of the best places in the world to see large numbers of wild polar bears in relatively close proximity (Dawson et al., 2010; Lemelin, 2006). Polar bears in this region stay on land when the sea ice thaws in summer and tourism operators run polar bear tours starting in July. Visitors arrive in droves during peak polar bear season, which lasts for approximately six weeks in October to November, when the ice begins to form on the Hudson Bay (Dawson et al., 2010).

Churchill's unique geography is the reason polar bears congregate here annually. Situated near the mouth of the Churchill River, it experiences earlier ice formation as the fresh water enters the ocean bay, which means that polar bears have historically congregated in this region waiting for the sea ice to form. This is also why the region has been inhabited, first by Dene and Cree peoples. The Churchill River was also a driving force in establishing the region as an historically important trade route, which eventually led to the building of railway infrastructure in order to more effectively transport grain to the port for shipment to international markets. Over the course of the last century a military presence helped to establish an airport in the region, as well as supporting infrastructure for the modern town of Churchill. Since there is no road access to this part of the province, rail and airport infrastructure was a necessary precursor to polar bear tourism development in the region.

These combined factors helped to propel Churchill, onto the international stage as a destination that is relatively accessible to tourist activity and promises a high degree of certainty of wildlife viewing opportunities. Tourists also come to see other animals, such as beluga whales and migratory birds at other times of the year, and the local tourism economy has developed enough to support several hotels and alternative accommodations that offer learning tours. The majority of visitors during bear season go on day-trip excursions on large, slow-moving tundra vehicles that allow them to see polar bears up close while remaining safely out of the polar bears' reach (Lemelin et al., 2010). As some polar bear tourists come to Churchill just for the day, while others plan longer trips, accurate numbers are not available, though it is estimated that between 6,000 and 10,000 travel to Churchill each season (Dawson et al., 2010).

Churchill is not the only location gaining popularity as a place to see polar bears in the province. The Assiniboine Park Zoo in Winnipeg can house between seven and ten polar bears in its "Journey to Churchill" exhibit and the Leatherdale International Polar Bear Conservation Centre at any one time. The "Journey to Churchill" (JTC) exhibit at the Assiniboine Park Zoo opened in the summer of 2014, following many years of planning and construction. The exhibit includes a variety of Arctic species of mammals and birds, such as polar bears, seals, muskoxen, wolves, and snowy owls, as well as interpretive elements including a representation of the town of Churchill. Interpretive elements feature exhibits about climate change, and the Aurora Borealis Theater runs a short interpretive movie from an Indigenous perspective. The exhibit also includes The Leatherdale International Polar Bear Conservation Centre, which functions as an interpretive space in the area accessible to visitors, and as an active research lab and transition center for orphaned polar bear cubs. With increasing frequency polar bear cubs are found orphaned in the Churchill region. Polar bear scientists attribute this to be, in part, due to the increasing effects of climate change – where mothers might abandon or be separated from their cubs while waiting for the sea ice to form (Derocher & Stirling, 1996). For years, when sea ice thaw has been especially early or freeze-up has been delayed, higher incidences of abandoned cubs have been observed (Nelitz et al., 2015). Polar bear cubs that are orphaned in the wild under the age of two have little chance of survival and whether to leave them to perish or to "rescue" them is intensely debated in Churchill.

When orphaned cubs are identified around Churchill, they are assessed by the Assiniboine Park Zoo's veterinary team for their suitability to be transferred to the conservation center, where they become acclimatized to the greater presence of people and are given any additional care. Eventually these polar bears are transitioned into the main JTC exhibit and may later be moved to other accredited zoos. However, under provincial legislation the polar bears are considered property of the government of Manitoba in perpetuity, which affords the province the ability to track and regulate the care of polar bears (Government of Manitoba, 2019). As these governmental processes and legislation demonstrate, the polar bear has become entangled in the relationship between the human species and its environment and subjugated to the effects of a warming climate and loss of habitat. This has simultaneously enabled development of these sites of polar bear tourism and has exacerbated the effects of climate change. Concomitantly, through the dichotomizing of human landscapes and nature and wildlife, the entangled past of wildlife tourism has subjugated polar bears to the work as an ambassador both for their own species and for the well-being of the planet.

Polar bears: Ambassadors for conservation and climate change

Polar bears in Manitoba work as ambassadors in two ways. First, wild polar bears work generally as ambassadors for climate change (Manzo, 2010). As a species, they are symbolically associated with climate change and are employed by the media to evoke emotional connection and empathy for climate change related issues (Born, 2019; Manzo, 2010). Their work extends beyond the Arctic regions where they live, crossing international boundaries as a symbol of climate change for conservation organizations and they are featured prominently in media headlines (Born, 2019). For example, Coca Cola uses polar bears in their advertising and also to partner with climate change initiatives; similarly, the polar bear as a symbol of climate change is utilized by trusted authorities such as National Geographic (Born, 2019). The icon of the polar bear entails a certain gaze of its own and may in turn influence how tourists choose to photograph and memorialize their experiences (Born, 2019).

Ex situ, or captive, polar bears in Manitoba serve as "ambassadors" for their wild counterparts (Assiniboine Park Zoo, 2020). Here, these bears act as ambassadors for their species, helping to educate visitors about life in the north and climate change issues. The Assiniboine Park Zoo's polar bears exist, in part, to educate the public and create an emotional connection between urban and southern Manitobans and their northern relatives. Indeed, the creation of the "Journey to Churchill" Exhibit was a response to the Polar Bear Protection Act which was passed in 2002 (Government of Manitoba, 2019), and facilitated the transfer of abandoned polar bear cubs to the Assiniboine Park Zoo for conservation and education. Ascribing to captive bears the role of ambassadors for their wild counterparts and raising awareness of climate change, is used to justify this captivity (Assiniboine Park Zoo, 2020). Further, conferring the title of "ambassador" implies the expectation that polar bears accept this elevated position.

In relation to moral ordering and conferring roles and social values to animals, Kline (2018) poses the following question with regard to whale watching:

> Perhaps there is social value in asking ourselves when and where else do we engage in moral ordering when it comes to watching vs. eating vs. some other use? Where else do our research efforts need to be more nuanced and holistic? (p. 214)

We argue that this exploration of the social value and the use of wildlife in tourism needs to extend beyond consumption, whether literal or visual. Exemplified in polar bear tourism, the moral ordering of polar bears as ambassador species is bound up in their work in climate change and conservation narratives. The use of wild polar bears is commodified in the form of tourism, particularly last chance tourism (Lemelin et al., 2010), and social in the use of polar bears as an emotional symbol for helping people to connect with and understand the severity and urgency of climate change (Born, 2019). Wild polar bears are situated in an anthropocentric moral ordering where they are important because they help the human species to recognize the harm we are continuing to cause to the planet and other species, and of course ourselves (Huijbens & Einarsson, 2018).

Ex situ, or captive, polar bears are used in a similar manner within the zoo context, where climate change messages are also employed via the medium of the polar bear to help visitors connect with and learn to care about climate change (Bueddefeld, 2019). Yet, captive polar bears are situated further down the moral ordering, as they also work as ambassadors for their wild counterparts. They work in the zoo, as part of a tourist attraction, as a lively commodity, which works to educate and entertain and to support its own existence by attracting visitors in order to pay for the high cost of housing polar bears in a captive setting (Figures 12.2 and 12.3). Furthermore, these captive bears have also paid the price of human impact on the landscape.

Central to understanding working animals is species relations of power and how the relative power held by different animal groups is expressed in their circumstance and experiences (Hovorka, 2019). Not only are animals positioned relative to humans, but they are also positioned relative to each other. The labor of wild and captive polar bears is valued differently on the basis of how wild these groups are perceived to be, and how these perceptions are mediated by their entanglement with human-animal systems (Collard, Dempsey, & Sundberg 2015; Dashper, Chapter 2). The labor of wild and captive polar bears is closely linked with their relationships to the broader ecosystem and its response to climate change.

Further, Braverman (2017) uses a "zoometric" hierarchic scale to compare the relative importance of animals, and in this context the polar bear in its natural ecosystem is valued in terms of being the top predator. However, in terms of perceptions of wild versus captive polar bears, wild bears are viewed as more authentic and therefore with greater value (Bueddefeld, 2019). These perceptions also vary between cultures in terms of the subjectivities that may be ascribed to polar bears. Traditional Inuit hunters believe that polar bears are willing participants in the

Figure 12.2: Children watching a polar bear at an ex situ zoo.
Source: Unsplash (Open source). Photo by Vidar Nordli-Mathisen.

hunt and that the imposition of a quota, to conserve populations, would break their trust with polar bear (Lemelin & Maher, 2009); this led to concerns by the Inuit that the polar bear would view them as arrogant and to fears of retribution. On the other hand, the Western view of animals as objects saw the polar bear as augmenting the lived experience of humans and being "a species, a living resource, or a national treasure" (Boyer, 2011, p. 106). These cultural differences in the polar bear gaze highlight the distinction between recognizing the polar bear as an autonomous agent, acting in its own narrative, as opposed to it being an actor in a human-centric narrative.

Neoliberal natures and green governmentality

Critical animal studies trace the moral ordering and neoliberal processes which have made whales (Huijbens & Einarsson, 2018), pigs (Zhou & Grimwood, 2018), elephants (Duffy, 2014), and other animals into lively commodities (Collard, 2013) to be consumed and used by the human species as physical resources in their death and as experiential touristic resources in their life (Rutherford, 2011). The term "lively

Figure 12.3: Visitors on a tundra vehicle photographing a polar bear near Churchill, Manitoba. Source: Author.

commodity," here, refers to living beings whose value is entangled with capitalist processes that render them commodities both in life and death (Collard & Dempsey, 2013; Collard et al., 2015; Rutherford, 2011). Previous research has chronicled the ways in which animal bodies became lively commodities through an array of complex histories: the exotic pet industry (Collard, 2013), animals working as physical laborers and tourism laborers in Thailand (Duffy, 2014), and the process of using captive elephants in Botswana for tourism (Duffy & Moore, 2010). Similarly, research examining shifts in the way animals such as whales have been used – from a resource in whaling to a resource in tourism – identifies a tangible shift in the moral ordering of whales (Huijbens & Einarsson, 2018) as a lively commodity in neoliberal "green" economies (Collard, 2014). Critical theory approaches have made visible the moral ordering of wildlife tourism and have furthered our understanding of how animals are brokered in a complex system of commodities and are commodified for entertainment or education (Collard, 2014; Collard & Dempsey, 2013; Derrida & Wills, 2002; Duffy, 2014; Rutherford, 2011).

Current tourism practices consume wildlife as lively commodities in such a way that perpetuates governmentality; more specifically, a green governmentality where there are moral arguments made to support the use of particular animals as lively commodities (Collard, 2013; Duffy, 2014). Green governmentality addresses how polar bears have been made to work as lively commodities in education, as ambassadors for conservation and sustainability narratives. In research exploring the discursive role of Giant Pandas in politics, Collard (2013) demonstrates that the panda's work is not benign and merely symbolic, but has geopolitical and material implications: "A symbol is far too static a role to account for the dynamic, un-predictable, and material effect that animals can have in political and environmental struggles" (p. 228). Similarly, polar bears as ambassadors transcend the role of static symbol into a material and lively commodity both in tourism and within the climate change discourse.

Tourism, in particular nature-based tourism, is touted as a mutually beneficial industry, where economic profits are in a symbiotic relationship with local people and culture, which promotes the conservation of wildlife and ecosystems (Duffy, 2015). Indeed, green capitalism and the greening of the tourism industry has become integral to sustainable tourism strategies promoted by the United Nations (Duffy, 2015). Yet, the contradiction between unrestricted economic growth and the support of conservation efforts seems to be largely overlooked. Succinctly described by Duffy (2015, p. 531): "Wanner (2014) argues that the green economy is a means by which neoliberal forms of capitalism sidestep and obfuscate the contradictions inherent in the tensions between continual economic growth and ecological limits." The virtue instilled in nature-based tourism, especially where polar bears are employed to teach people about climate change, obscures the fact that the neoliberal processes of tourism in ecologically remote locations such as the Arctic contributes directly to climate change through the carbon emissions associated with travel (Dawson et al., 2010; Lemelin et al., 2010).

Polar bears as lively green commodities

As demonstrated in these case studies, polar bears make visible the tensions between economic growth and ecological limits. For visitors going to see polar bears in Churchill, a moral argument is made whereby a visit to an ecologically sensitive area, requiring a high carbon output, is justified by the value in seeing the polar bear as an ambassador for climate change (Dawson et al., 2010; Lemelin et al., 2010). This justification is often compounded by the increasing market value of last chance tourism opportunities, where the value of the polar bear as a lively green commodity increases in correlation to the negative impact on its habitat and likelihood of long-term survival (Lemelin et al., 2010). Green governmentality is the control of social relations with nature, the natural world, and the environment through economic and political systems. The conception of polar bears as symbolic animals, which have values ascribed to them other than their individual value as living beings, relates to the power relationships and moral ordering that continues to shape wildlife tourism and underscores human relationships with other non-human species (Collard et al., 2015; Hovorka, 2019). How these tourism experiences are created and narrated has important implications for understanding the ramifications of wildlife tourism and the work of polar bears as lively green commodities.

Understanding polar bears as lively green commodities lays visible the practices of green governmentality, under which solutions are justified by framing the polar bear as an ambassador species; however, the work that the polar bear is made to do, or why the polar bear does this work as an ambassador species, is often left unquestioned (Born, 2019). How polar bears are conceptualized as lively green commodities where green governmentality frames the narrative has important implications for the critical thinking about the moral ordering of polar bears and nature in general. Examining the ethical implications of employing polar bears as ambassadors allows for a re-evaluation of the processes by which the human-animal relationships, as such, have formed and continue to reinforce the bears' identity as lively commodities in this touristic exchange.

The ethical implications in these examples are far reaching. As climate change worsens, additional polar bear cubs will likely be found abandoned. This will result in an increase in captive polar bears, and, hence, an increase in polar bear ambassadors at the zoo, and, ironically, more polar bears doing the work to save their wild counterparts. Thus, as the species continues to decline, their rarity perpetuates the polar bears' role as climate change ambassadors in the wild and increases their value as lively green commodities in last chance tourism. In other words, the polar bears' continued suffering through climate change positively benefits their symbolic and therefore economic value. Yet, as tourism to see the wild polar bears increases, so too does the carbon footprint and the effects of climate change, in an ever-worsening cycle of last chance tourism where the polar bear is perpetually working to the death of the species (Dawson et al., 2010; Lemelin et al., 2010).

The gaze is an approach for seeing the ordering and dichotomizing of nature in these tourism contexts (Lemelin, 2006). The power dynamics inherent in the processes by which polar bears are rendered lively green commodities can be better understood through the visitor's gaze of wildlife. Polar bears are made to work as ambassadors in both in situ and ex situ sites, in part, because of visitors and their desire to see polar bears. These experiences, then, may include narratives which frame the polar bears as ambassadors, with the hope of inspiring visitors to care about the species and take action in relation to their impact on climate change. Yet, without exploring the power dynamics which have created these conditions in the first place, as they relate to human and non-human relationships, we fail to critically think about the process of making decisions regarding the value of nature as it pertains to the impact on humans (Carr, 2016; Collard & Dempsey, 2013; Derrida & Wills, 2002). In short, with the additional lens of the gaze we make visible the power dynamics, or moral ordering processes, which have shaped relationships between humans and wildlife over the course of history and have become increasingly complex in modern tourism (Carr, 2016). This provides an additional lens of analysis that allows us to see beyond the literal or figurative consumption of animals through food or photography-based tourism, to their consumption as lively green commodities where tourists consume an ethical narrative of conservation and climate change.

Conclusion

This chapter has examined the work that polar bears are made to do as ambassadors, both for their own species and as a symbol of climate change. In exploring the context of these two case studies we have contrasted the ways in which wild and captive polar bears do their work as lively green commodities in climate change and last chance tourism narratives. Through a critical theory approach, the processes of green governmentality in the polar bear tourism industry are made visible (Rutherford, 2011). Green governmentality mediates social-environment interactions to the extent that the political and economic contexts that have created the current environmental crisis remain largely unquestioned. Moreover, while green governmentality seems to encourage behaviors that have environmental benefits, they are more likely to ensure the maintenance of the capitalist status quo. As Duffy (2015) explains, "In essence, nature-based tourism allows neoliberalism to turn the very crises it has created into new sources of accumulation" (p. 529). Hereby, the use of polar bears for their own conservation, and framing them within the climate change narrative, reinforces the visitor's gaze and the processes of green governmentality that make polar bears continue to labor as lively commodities. This is a process that will continue to the demise of the species, despite its green overtones. If polar bears are to do the work as conservation and climate change ambassadors, the next step in investigating the ways animals work in the tourism industry must be extended beyond their symbolic roles, which render them as lively green commodities and perpetuate the same system that created the current climate crisis. To begin to address and create "appropriate solutions" (Rutherford, 2011, p. 275) humanity must re-examine its current moral ordering (Kline, 2018), which is reinforced and perpetuated by the visitor gaze, rendering polar bears as ambassadors in the first place.

References

Assiniboine Park Zoo. (2020). *Polar bear rescue & care team*. Retrieved March 7, 2019, from https://www.assiniboinepark.ca/zoo/home/conserve/polar-bear-rescue-care-team

Berger, J. (1977). *Why we look at animals. About Looking*. New York: Pantheon.

Boyer, K. (2011). Polar Bears and Other Objects: Contextualizing the Failures in Saving Animals. In *Critical Norths: Space, Nature, Theory*, (pp. 103–117). University of Alaska Press.

Braverman, I. (2011). Looking at zoos. *Cultural Studies, 25*(6), 809–842.

Braverman, I. (2012). *Zooland: The institution of captivity*. Stanford University Press.

Braverman, I. (2017). Captive: zoometric operations in Gaza. *Public Culture, 29* (1(81)), 191–215.
Born, D. (2019). Bearing witness? Polar bears as icons for climate change communication in National Geographic. *Environmental Communication, 13*(5), 649–663.
Bueddefeld, J. (2019). *Exploring nature-based tourism and visitor learning at in situ and ex situ destinations*. [Doctoral dissertation, University of Manitoba]. MSpace.
Carr, N. (Ed.). (2016). *Domestic animals and leisure*. Springer.
Cronon, W. (1996). The trouble with wilderness: or, getting back to the wrong nature. *Environmental history, 1*(1), 7–28.
Collard, R. C. (2013). Panda politics. *The Canadian Geographer/Le Géographe canadien, 57*(2), 226–232.
Collard, R. C. (2014). Putting animals back together, taking commodities apart. *Annals of the Association of American Geographers, 104*(1), 151–165.
Collard, R. C. & Dempsey, J. (2013). Life for sale? The politics of lively commodities. *Environment and Planning A, 45*(11), 2682–2699.
Collard, R. C., Dempsey, J., & Sundberg, J. (2015). A manifesto for abundant futures. *Annals of the Association of American Geographers, 105*(2), 322–330.
Dawson, J., Stewart, E. J., Lemelin, H., & Scott, D. (2010). The carbon cost of polar bear viewing tourism in Churchill, Canada. *Journal of Sustainable Tourism, 18*(3), 319–336.
Derocher, A. E. & Stirling, I. (1996). Aspects of survival in juvenile polar bears. *Canadian Journal of Zoology, 74*(7), 1246–1252.
Derrida, J. & Wills, D. (2002). The animal that therefore I am (more to follow). *Critical inquiry, 28*(2), 369–418.
Duffy, R. (2000). Shadow players: Ecotourism development, corruption and state politics in Belize. *Third World Quarterly, 21*(3), 549–565.
Duffy, R. (2014). Interactive elephants: Nature, tourism and neoliberalism. *Annals of Tourism Research, 44*, 88–101.
Duffy, R. (2015). Nature-based tourism and neoliberalism: Concealing contradictions. *Tourism Geographies, 17*(4), 529–543.
Duffy, R., & Moore, L. (2010). Neoliberalising nature? Elephant-back tourism in Thailand and Botswana. *Antipode, 42*(3), 742–766.
Fennell, D. A. (2011). *Tourism and animal ethics*. Routledge.
Foucault, M. (2008). *The birth of biopolitics: lectures at the Collège de France, 1978–1979*. Basingstoke, Hampshire, GBR: Palgrave Macmillan.
Franklin, A. (1999). *Animals and modern cultures: A sociology of human-animal relations in modernity*. Sage.
Ginsberg, J. R. (1993). Can we build an ark? *Trends in ecology & evolution, 8*(1), 4–6.
Government of Manitoba. (2019). *Manitoba laws: The Endangered Species and Ecosystems Act*. Retrieved March 7, 2019, from http://web2.gov.mb.ca/laws/statutes/ccsm/e111e.php
Hanson, E. (2004). *Animal attractions: Nature on display in American zoos*. Princeton University Press.
Hallman, B. C. & Benbow, S. M. P. (2007). Family leisure, family photography and zoos: Exploring the emotional geographies of families. *Social & Cultural Geography, 8*(6), 871–888.
Hovorka, A. J. (2019). Animal geographies III: Species relations of power. *Progress in human geography, 43*(4), 749–757.
Huijbens, E. H., & Einarsson, N. (2018). Feasting on friends: Whales, puffins, and tourism in Iceland. In *Tourism Experiences and Animal Consumption* (pp. 10–27). Routledge.
Kline, C. (Ed.). (2018). *Tourism experiences and animal consumption: Contested values, morality and ethics*. Routledge.

Lemelin, R. H. (2006). The gawk, the glance, and the gaze: Ocular consumption and polar bear tourism in Churchill, Manitoba, Canada. Current Issues in Tourism, 9(6), 516–534.

Lemelin, H., Dawson, J., Stewart, E. J., Maher, P., & Lueck, M. (2010). Last-chance tourism: The boom, doom, and gloom of visiting vanishing destinations. *Current Issues in Tourism, 13*(5), 477–493.

Lemelin, H. & Maher, P. (2009). Nanuk of the Torngats: human–polar bear interactions in the Torngat mountains National Park, Newfoundland and Labrador, Canada. *Human Dimensions of Wildlife, 14*(2), 152–155.

Lemelin, R. H., & Wiersma, E. C. (2007). Perceptions of polar bear tourists: A qualitative analysis. *Human Dimensions of Wildlife, 12*(1), 45–52.

Markwell, K. (Ed.). (2015). *Animals and tourism: Understanding diverse relationships* (Vol. 67). Channel View Publications.

Manzo, K. (2010). Beyond polar bears? Re-envisioning climate change. *Meteorological Applications, 17*(2), 196–208.

Mason, P. (2000). Zoo tourism: The need for more research. *Journal of sustainable tourism, 8*(4), 333–339.

Moss, A., Esson, M., & Bazley, S. (2010). Applied research and zoo education: The evolution and evaluation of a public talks program using unobtrusive video recording of visitor behavior. *Visitor Studies, 13*(1), 23–40.

Nelitz, M. C., Wedeles, R. H., Lemelin. R. H., Beardmore, B., & Abraham, D. (2015). *Tourism carrying capacity review of the Churchill wildlife management area: Summary, key findings, and recommendations.* Report prepared for Manitoba Conservation and Water Stewardship, Winnipeg, MB by ESSA Technologies Ltd., Vancouver, BC.

Packer, J. & Ballantyne, R. (2012). Comparing captive and non-captive wildlife tourism. *Annals of tourism research, 39*(2), 1242–1245.

Packer, J. & Ballantyne, R. (2016). Conceptualizing the visitor experience: A review of literature and development of a multifaceted model. *Visitor Studies, 19*(2), 128–143.

Rickly, J. M. (2018). Considering service animals in tourism. *Annals of Tourism Research, 71, 57–58.*

Roe, K., McConney, A., & Mansfield, C. F. (2014). The role of zoos in modern society – A comparison of zoos' reported priorities and what visitors believe they should be. *Anthrozoös, 27*(4), 529–541.

Rutherford, S. (2011). *Governing the wild: Ecotours of power.* U of Minnesota Press.

Stoinski, T. S., Allen, M. T., Bloomsmith, M. A., Forthman, D. L., & Maple, T. L. (2002). Educating zoo visitors about complex environmental issues: should we do it and how? *Curator: The Museum Journal, 45*(2), 129–143.

Tribe, A. & Booth, R. (2003). Assessing the role of zoos in wildlife conservation. *Human Dimensions of Wildlife, 8*(1), 65–74.

Uddin, L. (2015). *Zoo renewal: White flight and the animal ghetto.* U of Minnesota Press.

Urry, J. & Larsen, J. (2011). *The tourist gaze 3.0.* Sage.

Wanner, T. (2014). The new 'passive revolution' of the green economy and growth discourse: Maintaining the 'sustainable development' of neoliberal capitalism. New Political Economy, 20, 2141.

Whatmore, S. (2002). *Hybrid geographies: Natures cultures spaces.* Sage.

Young, J. & Carr, N. (Eds.). (2018). *Domestic animals, humans, and leisure: Rights, welfare, and wellbeing.* Routledge.

Yudina, O. & Grimwood, B. S. (2016). Situating the wildlife spectacle: Ecofeminism, representation, and polar bear tourism. *Journal of Sustainable Tourism, 24*(5), 715–734.

Zhou, T. & Grimwood, B. (2018). Consuming Shangri-la: Orientalism, tourism, and eating Tibetan savory pigs. In *Tourism Experiences and Animal Consumption* (pp. 28–41). Routledge.

Yulei Guo

13 “Cute, but get up and work!”: The biophilia hypothesis in tourists’ linguistic interactions with pandas

Introduction

The biophilia hypothesis explains that humans have an inherent preference to seek connections with other forms of life and with nature (Wilson, 1993). Scholars have recognized the significant roles that the biophilia hypothesis plays in human evolutionary history as the foundation upon which human beings build their relationship with the external world (e.g. Gullone, 2000; Heerwagen, 2009). According to Edward O. Wilson (1993), who established the biophilia hypothesis, the human mindset evolved in a biocentric world that prompted an awareness of community, focusing on living and cooperating with nature. While human beings have evolved into a machine-regulated world since the modern age, the human-nature relationship has been redefined as self-interest, rather than community awareness. It has been argued that a self-interest based human society is a disturbance to the human-nature harmony of pre-industrial society. Hence, a set of ethical codes of conservation is established to balance human beings’ relationship with nature, as Kellert (1993a, p. 60) observes: “an ethical responsibility for conserving nature stems, therefore, from more than altruistic sympathy or compassionate concern; it is driven by a profound sense of self-interest and biological imperative.” Conservation and conservation ethics, henceforth, are produced within the contemporary social relationships of self-interest. Kellert (1993b) and Wilson (1993) are observant of a paradox inherent in the machine-regulated world where the self-interest of human beings firstly leads to the exploitation and pollution of nature resources, but the same self-interest also simultaneously generates a need for conservation ethics.

Wilson’s (1993) biophilia hypothesis has been widely applied in formulating conservation ethics (e.g. Levy, 2003; Simaika & Samways, 2010) and furthering studies of ecology and biology (Besthorn & Saleebey, 2003; Kahn Jr, 1997; Van den Born, Lenders, de Groot, & Huijsman, 2001), but the theory also has a profound social aspect embodied in its entanglement with self-interest. This chapter provides a study of the relationship between self-interest and the biophilia hypothesis in wildlife tourism where the biological and societal forces are both present. According to Higginbottom (2004, p. 2), wildlife tourism is “tourism based on encounters with non-domesticated animals [. . . that] can occur in either the animals’ natural environment or in captivity.” The definition reveals that the biophilia hypothesis is contextualized in the tourism industry’s understanding of tourists’ self-interest that transforms biophilia interest into commodities. Higginbottom’s definition also

https://doi.org/10.1515/9783110664058-013

validates Kellert and Wilson's observation that self-interest initiates conservation ethics by adding a further example of wildlife tourism in which the self-interest driven market orients the biophilia hypothesis rather than the other way around.

This chapter takes the tourist-wildlife linguistic interaction zone as the empirical field to examine the dominance of self-interest of the tourists instead of the biophilia hypothesis. In particular, the researcher looks at how tourists *speak about*, *speak to*, and *speak as* the pandas in captivity modeled on Tannen's (2004) study of the owner-pet linguistic interactions in family context. Further, the research highlights tourists' expectations of pandas to be "working," which means being "cute" and performing "cuteness" through play and other activity – expectations that are communicated through these speaking modes. The linguistic interaction zone between tourists and wildlife has received scant attention from tourism and linguistic studies alike. Yet, the study shows that the field is rich with both empirical and theoretical implications.

Literature review

The biophilia hypothesis: A review

The importance of the biophilia hypothesis goes beyond the biological and genetic tendency of human beings to build strong affiliations with other species. Kellert (1993b) and Wilson (1993) point out that biophilia implies a human culture fashioned in the awareness of community in which individuals are interdependent members. The awareness of community shapes a human society that is integral to rather than separate from the rest of nature. Human kind, human ways of thinking and action, and human culture is but one of the multiple ways that different beings are connected to nature and nature's expressions. Being a human, therefore, is not necessarily the right, the divine, or the center of the world. The possibility to respond to the need for connection and become a member of a wider community satisfies the most profound yearning of human beings and creates meaning and depth within a human's life.

However, in the machine-world context, scholars have consistently challenged the influence of the biophilia hypothesis and its interpretative potential. For example, Beck and Katcher (2003) show the biophilia hypothesis has at least three limitations to address. First, because the Anthropocene has competed with biophilia continuously during the entire human history, the biophilia hypothesis does not underscore the concern for welfare and rights of animals as an inborn tendency of human beings. Second, biophilia, in terms of the preference for nature, life and life-like processes, is hard to verify and identify in both general and concrete contexts. Third, the cultural and biological influences are so intertwined that the biophilia hypothesis is impossible to see as a purely biological and genetic need.

The most significant gap that the biophilia hypothesis has to address is the translation from biophilia hypothesis to conservation ethics in modern context as self-interest is required to facilitate the transformation. Kellert (1993a) begins by categorizing the human-nature bond into nine different values – utilitarian, naturalistic, ecologistic-scientific, aesthetic, symbolic, humanistic, moralistic, dominionistic, and negativistic. Yet, the ethical responsibility for conservation, as Keller (1993a) observes, is driven by both self-interest and biological imperative. Kellert (1993a) shows the double nature of conservation as an ethical action – the altruistic compassion goes hand in hand with self-interest. There is, hence, a paradox located in the heart of the biophilia hypothesis that refers to a connected and harmonious community driven by the pursuit of self-interest of each individual. The task for the biophilia hypothesis is to reconcile the individualistic investment in self and the desire for a collective and harmonious community. In wildlife tourism, the love for animals has created both capital flows and grounded in a temporal human-wildlife encounter.

Wildlife tourism and biophilia hypothesis

The biophilia hypothesis underlying wildlife tourism means that tourists are motivated to encounter the life and lifelike process of the wild animals. Duffus and Dearden (1990, p. 215) argue that wildlife tourism belongs to the non-consumptive use of wildlife; namely, the sightseeing of wildlife is a “human recreational engagement with wildlife wherein the focal organism is not purposefully removed or permanently affected by the engagement.” Wildlife tourism experiences leave the wildlife intact and alive, making human-wildlife interaction a promising venue for sustainable consumption of nature. Scholars further link the non-consumptive wildlife tourism to the conservation of environment and nature (Burgin & Hardiman, 2015; Wilson & Tisdell, 2001). Ostensibly, biophilia hypothesis based wildlife tourism contributes directly to wildlife conservation (Higginbottom & Tribe, 2004). However, Duffus and Dearden (1990) are also aware that the non-consumptive use of wildlife is not driven by conservation ethics. Rather, wildlife conservation and the seemingly unharmed wildlife are an unexpected byproduct of the wildlife tourist’s self-interest in spending money for their touring experience.

The self-interest-sustaining wildlife tourism stems from the recognition of the benefits and psychological well-being that the encounter with wildlife generates. A growing number of studies have outlined such benefits. For example, Curtin (2009) notes the communication between tourists and wildlife enhances the psychological well-being of the tourists, leading to spiritual fulfillment and transcendental meaning making experiences. Ballantyne, Packer, and Sutherland (2011) observe that tourists build up sensory impressions, emotional affinity, and reflective and behavioral responses through wildlife encounters. Curtin and Kragh (2014) show that tourist-wildlife encounters help to address issues of urbanization and alienation in contemporary society by reawakening a "deeper ecological sub-consciousness" (p. 545). The individual benefits that each tourist can receive through their wildlife encounters drive the consumption of wildlife tourism but have also failed to form the community awareness that the biophilia hypothesis helped to generate in pre-industrial societies. Without community awareness, the biophilia hypothesis can be consumed individually – the tourist's meeting with the wildlife creates a meaning for his or her own life, his or her own spiritual enlightenment, or his or her own psychological satisfaction. The tourist's association with the biophilia hypothesis becomes a relationship between a consumer and the product rather than a member with his or her community, a change Karl Marx also elaborated with the concept of commodity fetishism.

The biophilia hypothesis in the linguistic contact zone

The linguistic contact zone (Figure 13.1) is a space enabling the non-consumptive exchange between tourists and wildlife facilitated by the tourists' language. The tourist's language also has its roots in the biophilia hypothesis and the self-interest of the tourists. The biophilia hypothesis based language provides the historical materiality and origin of human language itself, without which the tourists would

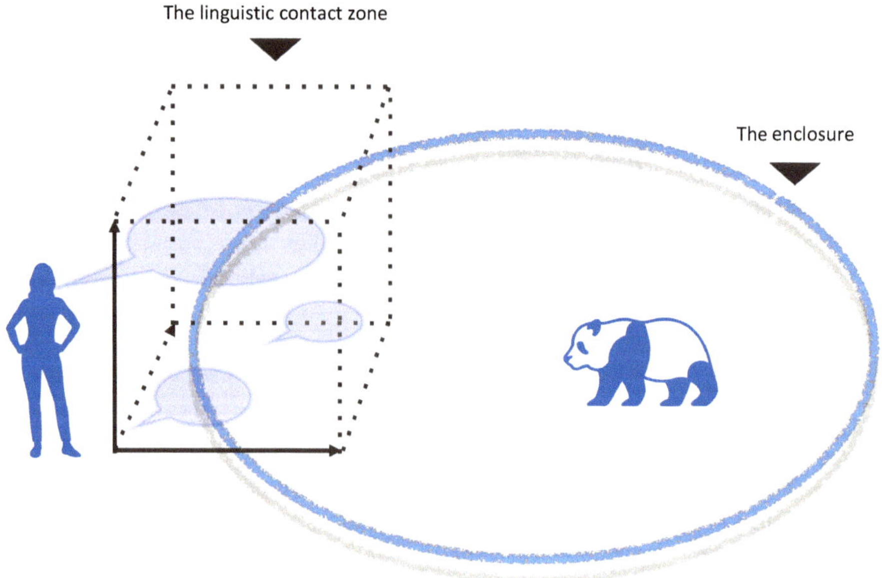

Figure 13.1: The linguistic contact zone.

have no ground for their linguistic interactions with the wildlife. As Kellert (1993a, p. 132) notes, "nature constituted the sole environment for our species' language development." The human language itself sustains the pre-industrial human-nature community awareness. Katcher and Wilkins (1993, p. 491) liken human language to the dance of bees – the language serves to indicate "the presence of plants or animals that was not visible." The language system in which each human individual has to learn makes him or her an integral part to a civil system that grows out from the human's interaction, observation, and preferences for the life and lifelike process.

The language of tourism, oriented by the tourists' self-interest, is rooted in the marketing and promotion materials that the tourism industry creates (Dann, 1996). Like a primitive human being who builds her knowledge by referencing nature, Dann (1996) points out that a contemporary tourist necessarily thinks, behaves, and talks through the tourism institution that connects her to the world. Also, because "people classify their own behavior in relation to that of others and use terms and categories in specific occasioned ways" (McCabe, 2005, p. 87) in tourists' narratives and conversations, it is understandable that being a tourist demands individuals to organize their language to address and elaborate their own self-interest. Despite the language's origin in the biophilia hypothesis, the language of tourism must weave self-interest and need of the tourists into the language they employ. Therefore, to understand how a tourist encounters a wild animal and obtains meaning from the encounter depends on knowing how the tourism system has referenced the wild animal. It does not mean that biophilia hypothesis-based human language system has collapsed. Rather, the engagement of the contemporary tourism industry suggests the need to consider the business as a mediator and variable in tourists' linguistic interactions with wildlife. The linguistic contact zone provides a rich empirical field for understanding the intersections between biophilia hypothesis and the self-interest of the tourists.

Talking to in wildlife tourism/zoo contexts

Though the linguistic contact zone is a new concept, researchers (Ballantyne, Packer, & Hughes, 2009; Higginbottom & Tribe, 2004; Mustika, Birtles, Everingham, & Marsh, 2013; Sinha, Qureshi, Uniyal, & Sen, 2012) have addressed the tendency of tourist/owners to converse with animals around them in a variety of contexts. These linguistic interactions stem from an anthropomorphizing tendency of human beings to look at other species that are, to use Arluke and Sander's (1996, p. 17) words, "regarded as a human in animal skin." Similarly, Patrick and Tunnicliffe (2012, p. 1) write, in zoo settings, tourists' initiated linguistic interactions with zoo animals become "an important part of understanding the knowledge gleaned by the visitors." Further, Mullan and Marvin (1999) suggest that many tourists in the zoo can only appreciate, understand, and feel for the animals by imputing human characteristics to the animals. It is argued, then, that the tourists' linguistic interactions with wildlife create a linguistic contact zone where the tourists' self-interest intersects with the biophilia hypothesis.

Other than wildlife tourism/zoo contexts, researchers have reported more results on the linguistic interactions between human beings, domesticated animals, and human infants without talking abilities. Arluke and Sanders (1996) recorded instances in which the pet owners speak for their pets. Shir-Vertesh (2012) shows that pet dogs are "flexible persons" or "emotional commodities" in families to which the owners communicate with varied underlying assumptions of the needs of pets and other family members. Burnhan, Kitamura, and Vollmer-Conna (2002) note that adults employ a special speech register, such as the elevated fundamental pitch, high affect, and exaggerated intonation contours, to address animals and human infants based on the adults' understanding of the emotional and linguistic needs of the audiences. These ways of talking to pets, it seems, are essentially biophilia because of the integration of the life and lifelike processes of the pets into the everyday life of the owners. As Tannen (2004) shows in her study, owners talk to, as, and about their pet in order to register the pet dog as a family member. The pet dog serves as a conversational resource that can help to exchange information and concerns, manage interpersonal interactions, and show affection and inclusion.

Pet animals serve as a conversational resource, allowing them to become a member to a household in the owners' everyday lives. What and how would a tourist talk to an animal stranger that he or she met in the zoo/wildlife context? If tourists in wildlife tourism and zoo contexts are communicating with the animals based on their self-interest rather than the biophilia hypothesis, how would the tourists conduct their conversations? In what way would the tourists address the wildlife? A closer look at the linguistic contact zone provides a better understanding to these questions.

Research design

The way to study the linguistic contact zone between tourists and wildlife is to observe the tourists' conversations with the wildlife and to record these conversations as accurately as possible. Using unobtrusive participant observation within the linguistic contact zone, the researcher was able to collect the natural flow of tourists' conversations with wildlife (Seaton, 1997). The research design is inspired by Tannen's (2010) study of the linguistic interactions between pet owners and their dogs. Tannen's (2010) team tracked two middle-class families' everyday conversations through contracted audiotaping for two weeks. In addition, they shadowed the participants for a day to observe the conversational contexts for the conversations.

The current study is similar to Tannen's project, as it shares a same interest in the conversations of people with animals. However, in this project the conversations taking place are in zoo and wildlife tourism settings and require a different data collection method. Seaton (1997) points out the unobtrusive, direct observations, also termed "mass observation," is a valuable addition to tourism research, though the research method has not been adequately applied in current tourism studies. This paper uses unobtrusive and mass observation to collect data within the linguistic contact zone, with a few adaptations to Tannen's study. First, the intensity and frequency of tourist-wildlife talks differ from the owner-pet conversations. The pets have a consistent presence in a family context, but the tourist-wildlife talks take place only when the wildlife is visible to the tourist. The owner-pet talks could extend over a day's time period, but the tourist-wildlife talks are specifically place-bounded with a higher level of intensity. The frequency and place-specificity of the tourist-wildlife linguistic contact zone allows the researcher to observe the research field on the site. Second, the tourist-wildlife talks take place in public spaces rather than in a family context where a contract is needed to protect the participants' privacy. The tourists talked to the wildlife in front of other tourists with a clear understanding that being a tourist is a public role rather than an individual and private identity. Nevertheless, the researcher chose to take field notes rather than audio or video recording the tourists as an ethical consideration.

Third, the research objective – the linguistic contact zone – suggests the researcher does not need to cultivate rapport with tourists in order to conduct the research. Instead, the study aims at understanding the zone and how the zone is used. The outdoor panda kindergarten next to the sun nursery house in Chengdu Research Base of Panda Breeding (CRBGPB) was selected for conducting the research. On June 5, 2019, Ji Li, the panda living in the Sun nursery house, gave birth to her daughter Ji Xiao. Two more panda cubs joined Ji Xiao the following month, making the Sun nursery house the most visited in the base and therefore the most frequently used linguistic contact zone. Having spent two months in the nursery, the panda cubs were strong enough to enjoy the fresh air in the outdoor kindergarten adjunct to the indoor nursery room. The research site was the outdoor kindergarten where the researcher could observe and record the talks of the tourists to the pandas (Figure 13.2). The enclosure standing between the panda cubs and tourist created a physical distance of 3–4 meters, measuring up to an invisible linguistic contact zone. In order to talk to the cubs, tourists raised the volume of their voices to cover the distance, making their talks publicly known. The researcher collected these talks between the tourists and the panda cubs (Figures 13.2 and 13.3). Three months of fieldwork for 1–1.5 hours observation each workday yielded 40-pages of notes, which the researcher further categorized and thematized on a weekly basis. The notes included the talks produced in the linguistic context zone and quick identification of the gender and age of the tourist as the researcher took glimpses of the tourists as the source of the talks. The records were initially coded into different themes on a daily basis and a weekly review helped to combine similar themes into unified and more general topics. After 3 weeks, the topics stabilized and three main

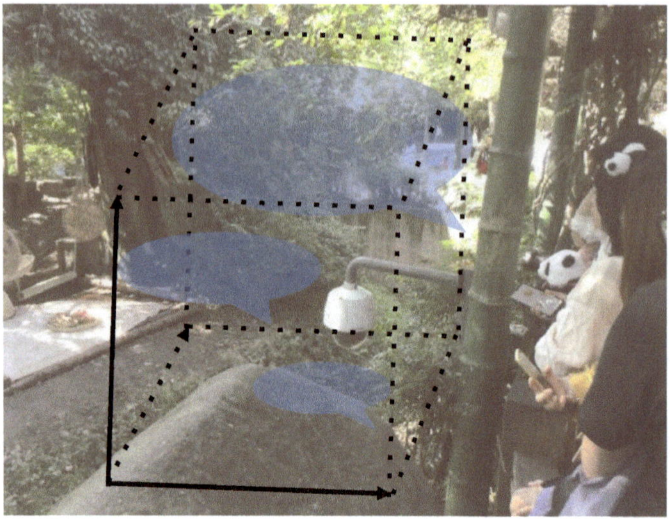

Figure 13.2: The linguistic contact zone visualized.
Source: Author.

Figure 13.3: A typical tourist crowds at the kindergarten enclosure.
Source: Author.

topics emerged. Further fieldwork yieldings were accordingly categorized under these topics. The researcher took notes on Mandarin, English, and other Chinese dialects she could understand. As one of the first studies on the linguistic contact zone, the project does not focus on knowing how varied language speakers have used the zone differently but provides a starting point for such research to follow.

Findings

The study modeled on Tannen's (2010) study of owner-pet talks, observed three ways that the tourists interacted with the pandas through the linguistic contact zone: speaking about, speaking to, and speaking as (Figure 13.4). The *speaking about* panda experience in the linguistic contact zone provides an image of the pandas that circulates among the tourists. Specifically, the study shows that talking about the panda's cuteness helps the tourist to identify with a global human community in which the panda's image provides an imaginative currency. The *speaking to* panda experience helps to clarify the assigned identity – the panda as worker – given to the pandas by the tourists. The direct conversations between the tourists and the panda are shaped by the social relationships structured within tourism business. In the *speaking as* the panda experience, tourists are most likely to enjoy the biophilia hypothesis but again the enjoyment depends on the panda's working performance. In sum, when tourists talk about pandas, they talk about "cuteness" as though being cute is the panda's job and part of their responsibility. As the tourists speaking as the pandas, they can appreciate the life and lifelike processes of the animal if pandas have performed their work decently.

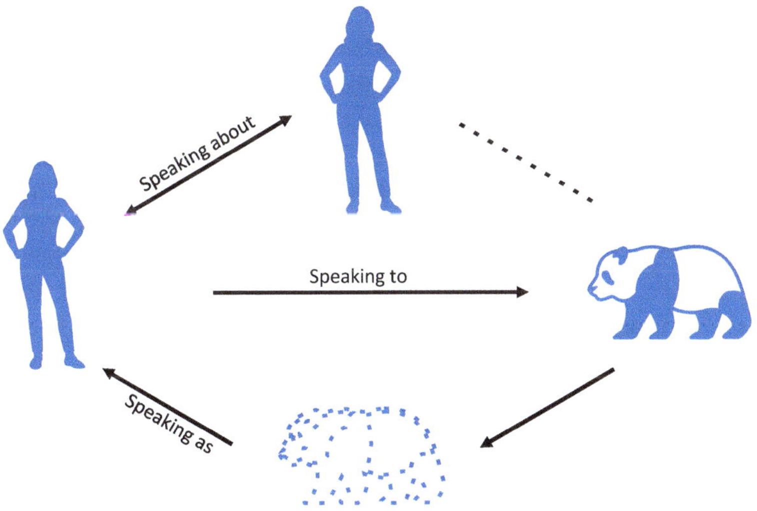

Figure 13.4: The three ways of tourist linguistic interactions.

Cute: What does a panda do?

The first word a tourist puts into the linguistic contact zone when meeting a panda is often "cute." The researcher recorded the word "cute" on a minute to minute basis. Cong et al. (2014) have a similar observation when they highlighted how tourists tend to employ the word "cute" to describe the pandas. Traditionally, "cuteness" has been explored from the biophilia hypothesis as scholars have been keen to explain human beings' natural emotional attachment to cute animals such puppies and kittens. Lawrence (1986) and Russow (1989) show that, as humans tend to form a stronger emotional bond with animals bearing youthful traits of human beings, the panda, who champions most of these traits, can cause people to react affectionately. The retention of the youthful traits of pandas, such as the round face, short chin and dark eye marks even into their adulthood, arouse the parental care in people (Morris & Morris, 1982). The tourists' exclamation of the pandas' cuteness can be explained as a natural reaction to humans' interest in the animal's human-like appearance and infantile traits.

Does it mean that, when tourists are talking about the cute pandas, they are invoking the biophilia hypothesis? Because the biophilia hypothesis is a genetic embodied preference, the recognition of the cuteness of pandas should start as a biological instinct. Yet, in the researcher's fieldwork, it was not uncommon to see that what is cute and recognized as cuteness is something that has to be learned and explained. Parents with young kids in their arms, pointing at the panda cubs, repeated the word "cute" to their children in an instructive manner. For the parents, the view of the pandas serves to exemplify "cute" and "cuteness" for their younger generation.

While scholars have shown that children tend to be better bonded with animals (e.g. Myers, 2007; Tunnicliffe, 1995), in the linguistic contact zone, the researcher observed that children do not necessarily bond with the pandas through the word "cuteness." Employing the expression of cuteness based on a well-defined understanding of what is cuteness demands a learned experience within the human culture itself. Therefore, it was interesting to see how young kids remained calm in front of the pandas, while adult tourists were eager to proclaim the "cuteness" of the pandas. The researcher observed the difference as she watched a boy identify the home objects of the pandas in the same way a man in his early 20s just did on the same site, while the boy did not relate the cubs to their cuteness. The difference between a boy and a grown-up tourist is not that the former could not understand what a panda is, but that the boy has yet to learn more about what cuteness is. The adult tourist saw the same pandas as the children did, but the tourists can easily anchor themselves in cuteness and its meaning.

Other than linking the cuteness of pandas to the biophilia hypothesis, there is an equally critical influence of tourism businesses and the tourists' self-interest that can help to create the image of a cute panda. Urry and Larsen (2011, p. 1) use the idea of the tourist gaze to explain that the tourism experience is "a learned ability and that the pure and innocent eye is a myth." Similarly, the notion of cuteness dwells in the linguistic contact zone between the tourist and the panda as a learned ability that the tourists have to acquire with a growing awareness of the idea of cuteness in the industry of tourism and the global culture. Talking about the cuteness of a panda offers an opportunity to communicate with a tourist community larger than the individual self standing in front of the animal's enclosure. It enables a communication between different tourists who tap into the linguistic contact zone as they agree upon the cuteness of the pandas. To a certain extent, to recognize the cuteness of the pandas is a part of being human in contemporary society. The self-interest does not allow the tourists to risk their human identity – claiming a panda ugly would be an anti-human act.

The construction of a cute panda meeting the tourist's self-interest has its own limitations that further enforces the recognition and dominance of human culture and community in the age of the Anthropocene. It leads to Nicholls's (2012) concern that there is no real panda in the global culture because of the animal's absolute presence in media and the cultural industry – the panda the tourists saw and talked was an image of panda created by the human culture and its global circulation. The cuteness talk allows the tourists to identify with a human global community and gain a human identity that might indirectly contribute to the cute animal's preservation and conservation. Yet, the self-interest could not push the tourist far beyond the imagined global image of the pandas and the centrality of human in the human-wildlife relationship, which can be best explained by studying how tourists talk to the pandas.

But you need to work

While the biophilia hypothesis can explain the tourists' preference for cute pandas, it cannot illuminate the tourists' direct demand that the pandas should "get up and work," a phrase frequently employed by the tourists when interacting with the pandas through the contact zone. In the researcher's observation, the tourists speak to the pandas as if the animals are workers who need to perform an essential job – being cute – for the tourists. The knowledge of meeting active and playing pandas circulates on TripAdvisor (Cong et al., 2014) as an important message exchanged between tourists, as they swarm to see pandas eating and playing in early morning hours. However, if talking about the cuteness indicates the tourists' belonging to a global culture community, the possibility to talk to the pandas has its root in the self-interest based social relationship that tourism businesses support – the tourist understands that his or her interest in psychological well-being and mental benefits depend on the panda's service. For the tourists to receive the expected value, the pandas need to work.

To a tourist's dismay, encountering a sleeping panda is inevitable as the pandas spend at 10 hours a day sleeping to maintain their low metabolism due to a vegetarian diet. Also, pandas are often inactive when the temperature rises during the day. Yet, for tourists, a napping panda is a "lesser" panda than a panda who is fighting, chasing, and playing with another panda. The linguistic contact zone has been utilized to make contact with the pandas that would return the pandas to their proper cuteness state. Thus, "pandas, get up. It is time to work!" is a phrase that many tourists resort to when seeing a sleeping panda. In the owner-pet study, Tannen (2010) show that owners speak to the dogs in order to include the non-human member into the family and to show the dog their affection. The pandas, however, have yet to gain a status that might enable tourists to perceive the pandas through a family bond.

However, the panda workers have yet to receive any job-related benefits, only responsibilities. Younger pandas have an even bigger responsibility to "work properly." In one observation, a mother came with her infant child in her arms. She soon joined other tourists to yell at the sleeping panda cubs to tell them it was time for work. After trying for a while, she decided that pandas were not going to wake up any time soon. She turned and walked away, telling her kid that the pandas were "off-work." She might not have realized that the panda cubs she was looking at and asking to do their jobs were newborn babies who were younger than the human baby she held in her arms. Yet, as young as the panda cubs are, the tourists have never doubted the cubs' work responsibility. Being the hottest spot in the base, the tourists were keen to observe a playful (working) panda cub.

Sleeping is not a proper way of doing one's job and tourists believe they have the right to wake the animals up. But the pandas' work responsibility goes far beyond the sleeping and non-sleeping dichotomy. For pandas who are not sleeping, tourists also instruct pandas to be playful and active. For example, the panda cub is at its best when it is playing or fighting with other cubs with their faces turned towards the cameras. A panda is also doing well when it does not venture into the bushes but instead stays on the most visible spot. During observation, the researcher noted a tourist persisted on making lip-smacking noises to "get the panda closer," while the proposal had been supported by a female tourist next to him who told him he was doing the right thing and then joined him. The speaking-to-panda interactions invoke a touristic social relationship that prioritizes the tourists' interest over the pandas' benefit. The tourism industry allows the tourists to exercise their power over those who have their job responsibilities, obliging the pandas to perform their jobs dutifully on a daily basis.

A good worker makes everyone happy

Identifying pandas as workers in the tourism relationship means the decent performance of the panda's job could please the tourist. Unlike pet dogs who can become a member to the family, pandas as workers imply a different social relationship. Tannen (2010) explored the owner's speaking-as-the-dog experience through the concept "constructed dialogue," which suggests the speaker animates the non-verbal participant's voice based on a shifted frame. The constructed dialogue assumes "a new and different footing vis-a-vis the participants and the subject of discourse" (Tannen, 2010, p. 402). The key to organize the constructed dialogue suggests the speaker can "adopt different roles" (Clift & Holt, 2006, p. 8) that "builds on and creates interpersonal involvement" (Tannen, 1986, p. 330). In the family context, the owners include the pet dogs as a significant member of the family. When owners shift into the dogs' footing, the owners were still speaking as a family member. A similar question needs to address the particular way tourists talked to the pandas. What does it mean that the tourist must shift into the panda's footing to construct interpersonal involvement with the animal?

The panda's footing is not as a family member, like the pet dog, but that of a worker. For the tourists, a worker does not speak as much as a family member does. The occasions that the tourists are speaking as pandas were limited during the researcher's observation. The tourists were keen to speak as the pandas when they observed the animals were actively engaged in their playing. The tourist dubbed the sounds of "Oh," "Ah," "Ouch," and "Hey" for pandas who were playing with each other or their toys. Researchers have argued that watching the pandas playing is critical to the tourists' satisfaction. Bexell et al. (2007) reported that tourists who have watched pandas playing demonstrated higher satisfaction about their travel experience. Cong et al. (2014) note that tourists tend to enjoy the scenes of active pandas more than inactive pandas. However, the researchers did not explain why watching the pandas playing turned out to be a satisfying experience for the tourists. Ostensibly, the biophilia hypothesis is sufficient to provide an explanation – the tourists were happier to see the pandas playing because they had the opportunity to observe the life experience better for the playful pandas than for sleeping pandas. Also, it is possible to argue that watching the playful pandas is an experience allowing the tourists to reconnect with nature because the experience benefits the well-being and psychology of the tourists. Watching lively pandas encourages the tourists to leave, temporarily, their urban landscape and everyday routine, to become a happy wild animal who sees no obligations and restrictions of the contemporary society (Curtin, 2009).

Yet, the biophilia hypothesis based explanation is, at its best, partial in explaining the tourist-panda relationship. The playful and active pandas mobilizing the tourists into a panda's footing are those who were doing a decent and diligent job. Identifying the pandas as workers means that shifting into a panda's footing requires the tourist to take on a worker's role. The "ah's" and "oh's" are exclamations in praise of the panda's excellent work, in parallel to the instructions and orders given to the sleeping pandas. The tourist does not appreciate the pandas as wildlife roaming freely in its own habitat, but they resonate with pandas who can be good workers. For urban tourists who have little experience of interacting with wildlife and wildlife established in former human-nature relationship, such as hunting, fishing, and collecting, the possibility for them to reflect on and recognize the existence of the pandas is to see the pandas as workers – a role the modern tourist is most familiar with. The possibility for tourists to appreciate the pandas as wildlife is very much out of the context in the light that such an appreciation requires a life experience in a different social context. Instead, tourists can appreciate the pandas through and within the social relationships they are living in – most tourists are not hunters, fishermen, or collectors anymore but workers who trade their labor for bread. The same working ethic is projected on pandas whose jobs responsibility is to be cute and to be a worker who performs cuteness. The fondness that the tourists bestowed upon the pandas does not necessarily mean that the pandas offer a link between the tourist and nature and visiting pandas offer an opportunity to reconnect with nature. Rather, the pandas, at least in the linguistic contact zone, are celebrated precisely because they are good workers to perform the idea of cuteness. The tourists have learned to appreciate the social relationship the pandas live within. The social relationship is not about being with nature but a human society structured by its established cultural, societal, and economic rules centering on the tourist's self-interest.

Conclusion

The chapter begins with the paradox of Wilson's (1993) biophilia hypothesis, which claims there is an inherently biological preference of human beings and wildlife conservation in a machine-regulated world. Wilson shows that the preference of life and lifelike process does not necessarily lead to actions of conservation but depends on the self-interest of the individuals. This chapter explores the relationship between self-interest and the biophilia hypothesis in the context of wildlife tourism. The empirical field for the study is the linguistic contact zone through which tourists speak about, speak as and speak to pandas. When the tourists were speaking about the pandas, they talk about cuteness as the job responsibility of the panda cubs. The tourists organize their conversations with the pandas by identifying the pandas as workers. When the pandas are doing their jobs of being cute properly, the tourists recognize their working ethic and resonate with the panda's diligence by switching onto the pandas' footing.

In current tourism studies, the biophilia hypothesis has been seen as the motivation behind tourists' interaction with wildlife – the preference for life and lifelike processes initiate tourists' interest in connecting with nature and the perceived psychological benefits and well-being as a consequence of interacting with the wildlife. However, this study shows that the influence of the biophilia hypothesis is contextualized within the tourists’ self-interest that is culturally established in an industrialized society. When meeting the pandas, the tourists have to identify the animal with the social relation they are living with in order to receive the promised benefits from the biophilia hypothesis. Therefore, the pandas become workers who can perform cuteness for wildlife tourists who enjoy the cuteness service. The panda, representing the wildness in the global culture today, have been assigned a working role that wildlife tourism has structured. Cuteness is the job responsibility of the pandas as they have performed the job better than any other animals on this planet.

References

Arluke, A. & Sanders, C. (1996). *Regarding animals*. Philadelphia: Temple University Press.

Ballantyne, R., Packer, J., & Hughes, K. (2009). Tourists’ support for conservation messages and sustainable management practices in wildlife tourism experiences. *Tourism Management, 30*(5), 658–664.

Ballantyne, R., Packer, J., & Sutherland, L. A. (2011). Visitors’ memories of wildlife tourism: Implications for the design of powerful interpretive experiences. *Tourism Management, 32*(4), 770–779.

Beck, A. M. & Katcher, A. H. (2003). Future directions in human-animal bond research. *American Behavioral Scientist, 47*(1), 79–93.

Besthorn, F. H. & Saleebey, D. (2003). Nature, genetics and the biophilia connection: Exploring linkages with social work values and practice. *Advances in Social Work, 4*(1), 1–18.

Bexell, S. M., Jarrett, O. S., Lan, L., Yan, H., Sandhaus, E. A., Zhang, Z., & Maple, T. L. (2007). Observing panda play: Implications for zoo programming and conservation efforts. *Curator: The Museum Journal, 50*(3), 287–297.

Burgin, S. & Hardiman, N. (2015). Effects of non-consumptive wildlife-oriented tourism on marine species and prospects for their sustainable management. *Journal of Environmental Management, 151*, 210–220.

Burnham, D., Kitamura, C., & Vollmer-Conna, U. (2002). What’s new, pussycat? On talking to babies and animals. *Science, 296*(5572), 1435–1435.

Clift, R. & Holt, E. (2006). Introduction. In E. Holt & R. Clift (Eds.), *Reporting talk: Reported speech in interaction* (Vol. 24) (pp.1–15). Cambridge: Cambridge University Press.

Cong, L., Wu, B., Morrison, A. M., Shu, H., & Wang, M. (2014). Analysis of wildlife tourism experiences with endangered species: An exploratory study of encounters with pandas in Chengdu, China. *Tourism Management, 40*, 300–310.

Curtin, S. (2009). Wildlife tourism: The intangible, psychological benefits of human–wildlife encounters. *Current Issues in Tourism, 12*(5–6), 451–474.

Curtin, S. & Kragh, G. (2014). Wildlife tourism: Reconnecting people with nature. *Human dimensions of wildlife, 19*(6), 545–554.

Dann, G. M. (1996). *The language of tourism: A sociolinguistic perspective*. Wallingford: CABI International.

Duffus, D. A. & Dearden, P. (1990). Non-consumptive wildlife-oriented recreation: A conceptual framework. *Biological Conservation, 53*(3), 213–231.

Gullone, E. (2000). The biophilia hypothesis and life in the 21st century: Increasing mental health or increasing pathology? *Journal of Happiness Studies, 1*(3), 293–322.

Heerwagen, J. (2009). Biophilia, health, and well-being. In: Campbell, Lindsay; Wiesen, Anne, eds. *Restorative commons: Creating health and well-being through urban landscapes.* Pennsylvania: USDA Forest Service.
Higginbottom, K. (2004). Wildlife tourism: An introduction. In K. Higginbottom (Ed.), *Wildlife tourism: Impacts, management and planning* (pp. 1–14). Gold Coast: Common Ground Publishing in association with the Cooperative Research Centre for Sustainable Tourism.
Higginbottom, K. & Tribe, A. (2004). Contributions of wildlife tourism to conservation. In K. Higginbottom (Ed.), *Wildlife tourism: Impacts, management and planning* (pp. 99–123). Gold Coast: Common Ground Publishing in association with the Cooperative Research Centre for Sustainable Tourism.
Kahn Jr, P. H. (1997). Developmental psychology and the biophilia hypothesis: Children's affiliation with nature. *Developmental Review*, *17*(1), 1–61.
Katcher, A., & Wilkins, G. (1993). Dialogue with animals: Its nature and culture. In K. Stephen & W. Edward (Eds.), The biophilia hypothesis. Washington: Island Press.
Kellert, S. (1993a). The biological basis for human values of nature. In S. Kellert & E. Wilson (Eds.), *The biophilia hypothesis*. Washington: Island Press.
Kellert, S. (1993b). Introduction. In S. Kellert & E. Wilson (Eds.), *The biophilia hypothesis.* Washington: Island Press.
Lawrence, E. A. (1986). Neoteny in American perceptions of animals. In R. J. Hoage (Ed.), *Perceptions of animals in American culture* (pp. 57–76). Washington: Smithsonian Institution.
Levy, S. S. (2003). The biophilia hypothesis and anthropocentric environmentalism. *Environmental Ethics*, *25*(3), 227–246.
McCabe, S. (2005). 'Who is a tourist?' A critical review. *Tourist Studies*, *5*(1), 85–106.
Morris, R., & Morris, D. (1982). *The panda*. London: Penguin Books.
Mullan, B., & Marvin, G. (1999). *Zoo culture*. Urbana: University of Illinois Press.
Mustika, P. L. K., Birtles, A., Everingham, Y., & Marsh, H. (2013). The human dimensions of wildlife tourism in a developing country: Watching spinner dolphins at Lovina, Bali, Indonesia. *Journal of Sustainable Tourism*, *21*(2), 229–251.
Myers, O. G. (2007). *The significance of children and animals: Social development and our connections to other species*. West Lafayette, IN: Purdue University Press.
Nicholls, H. (2012). *The way of the panda: The curious history of China's political animal.* London: Pegasus Books.
Patrick, P. G. & Tunnicliffe, S. D. (2012). *Zoo talk*. London: Springer Science & Business Media.
Russow, L.-M. (1989). Changing perceptions of animals: A philosophical view. In R.J.Hoage (Ed.), *Perceptions of animals in American culture* (pp. 25–39). Washington: Smithsonian Institution.
Seaton, A. (1997). Unobtrusive observational measures as a qualitative extension of visitor surveys at festivals and events: mass observation revisited. *Journal of Travel Research*, *35*(4), 25–30.
Shir-Vertesh, D. (2012). "Flexible personhood": Loving animals as family members in Israel. *American Anthropologist*, *114*(3), 420–432.
Simaika, J. P. & Samways, M. J. (2010). Biophilia as a universal ethic for conserving biodiversity. *Conservation Biology*, *24*(3), 903–906.
Sinha, B. C., Qureshi, Q., Uniyal, V., & Sen, S. (2012). Economics of wildlife tourism–contribution to livelihoods of communities around Kanha tiger reserve, India. *Journal of Ecotourism*, *11*(3), 207–218.
Tannen, D. (1986). Introducing constructed dialogue in Greek and American conversational and literary narrative. *Direct and Indirect Speech*, *31*, 311–332.
Tannen, D. (2010). Talking the dog: Framing pets as interactional resources in family discourse. *Research on Language and Social Interaction*, *37*(4), 399–420.

Tunnicliffe, S. D. (1995). *Talking about animals: Studies of young children visiting zoos, a museum and a farm*. (Unpublished doctoral dissertation), King’s College University of London, London.
Urry, J. & Larsen, J. (2011). *The tourist gaze 3.0*. London: SAGE.
Van den Born, R. J., Lenders, R. H., de Groot, W. T., & Huijsman, E. (2001). The new biophilia: an exploration of visions of nature in Western countries. *Environmental Conservation, 28*(1), 65–75.
Wilson, C. & Tisdell, C. (2001). Sea turtles as a non-consumptive tourism resource especially in Australia. *Tourism Management, 22*(3), 279–288.
Wilson, E. O. (1993). Biophilia and the conservation ethic. In S. Kellert & E. Wilson (Eds.), *The biophilia hypothesis*. Washington: Island Press.

Part 5: **Reflections**

Carol Kline, Erica von Essen, Johan Lindsjö, Adélaïde Fouache, Lara Tickle, Katherine Dashper, Erika Andersson Cederholm, Brent Lovelock, Anita Burkevica, Michael Jones, Mariana Reis Macieira, Jillian M. Rickly

14 Working animal research: An agenda for the future

Introduction

The relationship between humans and animals is long and varied, entangled in cultural meanings, national and regional symbolism, ecosystem benefits, emotional bonds, and a myriad of ways in which animal bodies are leveraged for labor, entertainment, education, testing, food, clothing, and other capitalistic pursuits. We sat down with scholars at the Instagranimal Symposium in Uppsala, Sweden, in August 2019 to discuss meanings of and a research agenda for working animals in tourism. The interdisciplinary gathering, a "symposium on the animal ethics and welfare challenges in animal-based tourism" (Von Essen, Lindsjö, & Fouache, 2020), was developed due to a series of contemporary animal tourism controversies (see for example Cohen & Fennell, 2016; Mkono, 2018; Wearing, & Jobberns, 2015), as well as the rise in power of social media as a source of both positive and negative change (Carder, Plese, Machado, Paterson, Matthews, McAnea, & D'Cruze, 2018; Mkono & Holder, 2019). Instagranimal brought together scholars and practitioners (e.g. Parken Zoo, World Animal Protection) from a variety of geographic locations and disciplines and fields, including animal welfare, animal and environmental ethics, ethology, species and environmental conservation, geography, environmental communication, tourism and leisure studies, sociology, philosophy and anthropology, veterinary medicine, and tourism, for a two-day event (please see Von Essen, Lindsjö, & Fouache, 2020 for represented disciplines and full proceedings). The timing of the symposium coincided with the development of this edited collection and symposium organizers von Essen and Lindsjö were kind enough to allow this discussion to ensue with willing participants.

Readers should note the timeliness of this discussion. It took place just five months before the first publicly reported cases of COVID-19 in January of 2020, while the analysis occurred in summer 2020. As we face uncertainty regarding the tourism industry and reassess our relationship with the animal world, this arguably only further drives home the relevance of this discussion.

https://doi.org/10.1515/9783110664058-014

Approach

The focus group was held on the final day of the symposium as an "add on" session after all of the formal presentations and workshop discussions had concluded. The focus group lasted one hour and ten minutes, was held in the same room as the symposium sessions, was guided by six questions developed by Jillian Rickly and Carol Kline, and was led by Kline. Excluding Kline, participants represented six different universities and one zoo. The recording of the discussion was transcribed and coded in two rounds; first by question and then subsequently, by themes that were revealed in the first round of coding. After the discussion summary was written, participant checks were performed by sending the summary to all participants involved and soliciting feedback. Minor revisions were made based upon their comments.

The discussion questions were:

- What comes to mind when you think about the labor that animals perform in the tourism industry?
- Do you think this is an important and or timely topic that needs addressing? Why?
- What do you think about the current state of research on working animals within the context of tourism?
- What needs more attention within working animals in the context of tourism, species, topics, methods, communities, destinations, whatever comes to mind, what needs more attention?
- How will we achieve doing this research with shrinking resources and other priorities?
- What makes you hopeful for the future of animal research within the tourism context?

Findings

The first two questions served to frame the topic, outlining the contours of animal labor and describing what fell within those boundaries. Participants explored meanings of work/labor, comparing types and dimensions of human versus animal work, applying "employee" issues to animal work, and adapting human labor theories to animal work. Additionally, it was important to delineate how work in tourism is unique.

Animal labor

A number of species engaged with tourism work were chronicled, as well as the nature of their role in the tourism experience; a sample of those mentioned are hunting dogs, horseback safari, petting zoos, animal selfies, animals who carry people or gear, performing animals, emotional support animals, dogs who render assistance to the visually impaired, as well as the myriad of species engaged in ecosystem services. The notion of direct or indirect labor and visible and invisible (hidden) labor took stage as participants acknowledged that animals, and nature in general, are overlooked or forgotten as they often fall into the background or setting of the tourism experience. Lara Tickle noted:

> We don't actually see how much we use animals today in our daily lives [. . .] From the shoes we wear to the bags we have to the food we eat, they are absolutely everywhere on us [. . .] Like a tree growing an apple, we pick it, that's free. That's nature. So an animal doing a job, that's free because it's nature. It's the same with any labor that isn't paid. It's just not thought of.

Indeed, animals "work" within their habitats. And while the term "ecosystem services" is universally understood, the consideration of these services as "work" by the various parts of the natural environment is not commonly accepted. The notion of service, labor, and work have different connotations within societal and cultural contexts, and its value often depends on who is doing it, what type of activity it entails, whether there is compensation involved, and if it is visible or hidden. To Tickle's point above, a hidden service that an animal provides which indirectly or even directly affects a tourism setting (such as seed distribution in a forest or being part of the visual landscape of an open plain) would not be considered as "work" because it could happen anyway regardless of the presence of tourism. It may be considered to have value (if we stop to really think about it) but not on the same level as for example the human tour guide interpreting the setting. Other indirect services animals provide include helping set the context of place, being a part of the regional food chain, within interspecies assistance symbiotics, and of course in cleaning the soil, water, and landscape by processing human-made pollutants and other animal carcasses. Michael Jones noted that because we expect animals to do ecosystem work in order for us to use the setting to our touristic advantage, "they are in true capitalist fashion expropriated and used as a resource for capital accumulation." Erica von Essen described these indirect forms of labor in tourism as

> not just as spectacle stage performers, but in the way that they kind of bring associations to place and also in the context of terraforming, or we would say that in re-wilding for example, you can think of animals as being kind of laborers. Because they're put there to perform certain ecosystem services to benefit other species or to shape rivers and so on. And that all contributes to place and there is a lot of animal labor that goes invisibly into that process.

Other types of labor discussed were symbolic, security, entertainment, transport of people or goods, emotional and therapy, and death labor (e.g. animals hunted, animals used as bait, animals killed in entertainment and sports, and animals slaughtered for food). Some work is within their nature while some is against their nature, but even when it is a "natural" behavior (e.g. digging, sniffing, running), the participants suggested that many animals do not have the option of refusing the task. We propose then that there may exist a continuum of agency to no agency, and that the animals with no agency are essentially slaves (von Essen). Sloths plucked from the forest to be relegated as props for animal selfies, for example, are taken from their home and/or trafficked just as human slaves are (Kline). Katherine Daspher wondered how does the slavery label applied to those animals who are "owned," such as dogs, horses, etc.; she noted that it is good that we are finally starting to ask these questions within tourism academe.

In addition to the continuum whereby some animals are considered more important or endearing than others (Fouache; Joy, 2010; von Essen), there exists a range of "remuneration" or "benefits" that animals receive for their labor. Some only get food and water, while others receive shelter, care, intellectual stimulation, and social and emotional interaction. Similar to human employment conditions and "job security," animals are recruited and can get sacked. If animals don't meet the requirements for the job (police dogs being too cuddly, for example), their fate varies from context to context. And further still, the compensation in retirement is better for some animals than others (a seeing eye dog who receives a new home vs. racehorse who is often slaughtered for parts); more pointedly, animals do not have a choice as to when or how they are retired (Andersson Cederholm; Fouache; Reis Macieira; Tickle). The emotional toil on animals of their labor is rarely considered, for example when a therapy animal or police dog is moved from one person to another (Andersson Cederholm) or horses or elephants are continuously exposed to new riders (Lindsjö). Relative to cognitive stimulation on the job, only within the past decades has the "enrichment" of captive animals been explored and implemented. And the "work" or "retirement" of farm animals is rarely prioritized (Fouache) except in the case of farm sanctuaries where rescued animals can live out their natural lives (Kline). Dashper noted that megafauna often receives priority attention over other smaller or less appealing types of animals.

When discussing working animals in tourism, participants in the focus group were reminded of labor within a Marxist system. Animals were likened to a cog in a machine, invisible as individuals and absorbed by capitalism as a resource. Therefore, when discussing the elements of worker "rights" for animals, many similarities to human worker rights emerged: agency, conditions, treatment, remuneration, and "earning" of leisure and retirement. Within a traditional Marxist pyramid depicting capitalism, the top is populated by royalty and government leaders, followed by religious persons one tier lower, then military, then a wide array of bourgeois citizens, who sit on the backs of the many workers at the bottom who toil to keep society running. In the historic presentation of the pyramid, animals are absent. However, with increased awareness of animal work, animals now occupy a space alongside or under those at the very bottom (Tickle). In the quest for economic gain, we take animals as resources and fit them into our capitalist system.

Certainly, there is an apt comparison between the invisible work of animals to human female work (Andersson Cederholm). It's always there but not regarded as "real" work, whereas the carrying or pulling animals, their work is visible and may be more highly regarded as they are valued for their strength and speed. There are traditional uses of animals as workers, e.g. elephants, cows, horses, in countries where their work is seen as cultural or heritage tourism, and they are acknowledged as workers. However, many animals likely do not have agency or positive working conditions and remuneration; "We exploit laborers as we exploit animals and resources. Just another parallel that I'm drawing between the different theoretical walls" (Tickle).

Brent Lovelock raised the question about the relative importance of animal welfare issues with human issues in tourism. He cited examples of child sex tourism or providing accommodations for individuals with varied abilities, adding climate change as an existential threat to both humans and animals. Johan Lindsjö responded that the welfare of both cannot be separated because we are all interlinked (for information on a related framework, see One Welfare https://www.onewelfareworld.org/). Animal work (for humans) and human work (for the animals) are intertwined (Jones; Lovelock; von Essen). One of the values of looking at animal work is to think more about human work as well (Dashper). However, it is a timely topic because even as animal work is often invisible or ignored, it is at the same time even more visible thanks to social media (Lindsjö).

And lastly, we must consider that tourism is a zone of indulgence and permissiveness where humans might be a little less scrupulous about the way animals are perceived and treated, and certainly culture plays into this differential (von Essen). Animals in economically developing countries may be viewed differently than in developed countries. Relatedly, some cultures welcome close relationships with animals, whereas others observe greater taboos. Tourism by its nature is a space where very different worldviews can become intertwined, and our different values around different species offers a new opportunity for tensions and misunderstandings as well as enlightenment and peace.

Current state of research

As participants reflected on the current state of research, all marveled at the lack of attention the topic has received in light of the prevalence of animal tourism around the globe. Dashper commented, "There is research about animals doing work, where they're not very often *described* as doing work. I think they're just explained in other ways. So thinking of them at work is a very recent thing." For examples, zoo animals work (see Guo, Chapter 13) and petting zoo animals work, but we don't tend to think of what they do as labor (Burkevica). Being bred for any type of consumption is work, and certainly female animal bodies are "worked" to the benefit of humans for the consumption of honey, eggs, milk, veal, and other animal products to eat, yet this is not considered work on the part of the animals; similarly male bodies are used for meat, fur, leather, wool, teeth, scales, paws, and other parts. So while there is research about animals who are working, it isn't labeled as such and the work itself (as work) is not the focus. Similarly, Reis noted that there are studies about the bulls bred for bullfighting, but not within the context of the "show" or entertainment itself (see Lopez & Venegas, Chapter 5).

The focus group participants suggested many areas where research on working animals might be expanded. We noted a theoretical divide between studying animals as individuals versus studying animals as a species (Burkevica; Lindsjö). Burkevica commented that the more we know about species, the better we can understand their relationship with work. Von Essen mentioned the growing interdisciplinary field of interspecies communication, for example, animals expressing dissent, preferences, permission.

Participants noted the need to learn more about why animal abuse is happening, borrowing from animal studies, ethics, and other disciplines, as well as the contexts in which it happens, why we can't do or haven't done anything about it, and which policies and regulations would be effective (Daschper; Lovelock). It would be useful to catalogue the type of work that animals are doing in tourism in different places around the world, in addition to the different types of animals engaged, and the conditions they're working in; from there, how can we improve the conditions and the welfare? Additionally, the role of multiculturalism must be explored as to what is allowed across many countries and why (von Essen). Lovelock cited unethical hunting practices and contexts around hunting endangered species as topics ripe for exploration. Jones suggested the development of a method for prioritizing the more egregious forms of animal welfare violations, while Burkevica and Jones called attention to the fact that abusing individual animals spilling over to population effects by affecting animal's offspring.

Regarding aquatic animals, there is very little research relative to tourism, which is surprising to von Essen as the ocean is one of the last remaining wild places on earth; Fouache noted that the "sea parks" research agenda has not been exhausted, including dolphinariums and all types of experiences getting up close to aquatic animals and/or aquatic animals in captivity, and would like to know what really happens to the animals once they are retired from their captive, performing work. Participants wondered why there is resistance to the notion that fish feel pain and stress; despite the fact that there is considerable research done to prove this, it is not being infused into common knowledge about the experiences of aquatic animals. Similarly, more attention needs to be given to animals with supposed lower levels of sentience, and not just megafauna (Lovelock; Tickle).

Regarding the consumer, additional research is needed into complacency and apathy in consumers; what other options do we give them other than "don't purchase it"?

There has been an increase among social media users to remove geotags of images taken in fragile ecosystems in order to "protect" them from overtourism and from accidentally tipping off poachers as to location. What kinds of resistance could be embedded to "protect" animal workers? Are there prosumption models that could be taken from consumer behavior and applied to tourism? The phenomenon of animal selfies must be researched more, from the standpoint of the consumer as well as the "handler" of the animals, and the animals themselves (see Tully & Carr, Chapter 6).

Burkevica was interested in the role of the working animal within the organization: what is the relationship there with the animal and the animal's co-workers: other animals, as well as humans and managers. According to Kline, Lovelock, and von Essen, additional topics that are not researched within tourism include:

- What benefits and detriments do pets experience from travel?
- How is the movement of animals around the world relative to the black market and to the pet trade connected to tourism?
- Are there negative effects on human communities by increasing the welfare of animals?

Messages of hope moving forward

Despite the lack of attention and research to the topic, the participants offered some reasons for hope. For example, we are already seeing increased awareness, pressure from consumers, and local capacity building surrounding animal welfare in general, and animal welfare in tourism. Burkevica has seen an evolution in the education of zoo attendees for example who:

> Five years ago, the average complaint was that "We don't see animals. They're too far away. Why did I have to pay so much money?" Today, actually, people come and leave a comment after their visit "Oh, I have learned a lot, and it was a really nice park, and it feels like that animals were well here. Oh, I didn't see any hippos but the keepers suggested to come back in the afternoon because then they're active." I see the hope in that because they actually respect the welfare of the animals today, not just maximizing their day, the animal species they have seen. So, I see a big hope in that, that it continues actually forward.

Consumers are becoming more particular about their experiences, and tour operators want to differentiate themselves by offering ethical tours. The power of social media (Mkono & Holder, 2018) provides hope for social pressure on behavior relative to animals, and the engaged younger generation, Generation Z (Post-Millennials) are galvanizing their collective efforts to protect the earth and animals when so many generations before have failed.

Focus group participants felt that positive outcomes can come from a publicized negative experience; the experience is exposed, and people are outraged to the point of demanding change. Because climate change is dire and killing off species, there is "hope" that it has gotten the attention of the public, just as in the same manner that negative media attention about an individual animal who suffered in some way can rally support for more stringent welfare protections. Certainly exchanges between people with understanding (such as at symposiums such as these) help advance our collective thinking on issues. Any greater knowledge leads to finding areas that we can all agree on. Additionally, because of the multiple effects of climate change, we are entering a period of great change which gives us a chance to redesign our experiences.

There is cause for hope in applying the philosophy of the three "Rs" (replacement, reduction, and refinement) historically used in animal laboratory experiments and testing also in animal-based tourism:

- *Replacement* of animals with alternative methods (e.g., activities without an animal theme, virtual reality)
- *Reduction* of number of used animals (e.g., less animals exposed in an enclosure or in a safari, or hunted for trophies)
- *Refinement* of methods and activities to minimize suffering and enhance animal welfare (e.g., good care, handling and environment, considerate and careful approach) (Lindsjö).

In brainstorming ways to effectualize and expand animal welfare research, we must educate the industry to understand the positive effects it will bring to them. Additionally, if the general public do not know about (or think about) these issues, then raising awareness is the first step; however, networking in specialist groups such as the symposium is equally critical in order to identify crucial next steps and develop critical mass among researchers and practitioners (Burkevica; Lindsjö).

An effective media relations strategy is to connect animal tourism issues with other issues already in the media (e.g. zoonotic disease, climate change, or political events; von Essen). Connecting animal issues with human issues also makes sense because we are co-mingled, but also some audiences are always going to prioritize human miseries over non-human ones, regardless of the relationships between the two. And tying in with human well-being would garner attention from the more anthropocentric among us.

Tapping into citizen science may be a way to extend efforts with limited funding for research, and certification schemes that give funding back to a species could help stretch research dollars (Jones; Reis). We might leverage the desire for research on this topic through the attention and interest given to the Sustainable Development Goals. Indeed, this research area has strong connections not only to SDG 14 Life on Land and SDG 15 Life Below Water, but also SDG 12 Responsible Consumption and Production and SDG 10 Reduced Inequalities. Working collectively to develop a solid research agenda involving key players from key areas and involving industry organizations that might play a critical role (UNWTO, WTTC) could help an animal welfare research agenda to gain traction (Lovelock). And creating opportunities for sponsorship through private sector funding, by getting industry partners on board. However, the greatest hope, if even unlikely would be that by continuing to educate all of our constituents (students, industry partners, communities, and each other) that sustainability in the long run includes investing in animal welfare and healthy environments.

Conclusion

This book project began with a kernel of an idea – what if we look at the animals of the tourism industry through the lens of work and labor? What would we find? There is already a strong literature documenting the inequalities and injustices of human labor in tourism; is animal labor comparable? If so, how do we move forward in scholarship and in the industry? Sitting down with impassioned animal researchers at the Instagranimal Symposium in Uppsala highlighted many of themes also raised in this collection of chapters, but also raised additional issues for investigation. The conversation also suggests we are approaching a turning point in tourism studies, where animal research is galvanizing an agenda and growing in strength. Within the introductory chapter, we stated "This starting point thus inspires a discussion of the relations of animal welfare, well-being, and their rights as workers." Indeed, we can observe a critical mass developing around research on these issues. Our colleagues' eyes are wide open and they are inspired (and inspiring).

References

Carder, G., Plese, T., Machado, F. C., Paterson, S., Matthews, N., McAnea, L., & D'Cruze, N. (2018). The Impact of 'Selfie 'Tourism on the Behaviour and Welfare of Brown-Throated Three-Toed Sloths. *Animals*, *8*(11), 216.

Cohen, E. & Fennell, D. (2016). The elimination of Marius, the giraffe: Humanitarian act or callous management decision?. *Tourism Recreation Research*, *41*(2), 168–176.

Joy, M. (2010). Why We Love Dogs. Eat Pigs, and Wear Cows: An Introduction to Carnism, 96.

Mkono, M. (2018). The age of digital activism in tourism: Evaluating the legacy and limitations of the Cecil anti-trophy hunting movement. *Journal of Sustainable Tourism*, *26*(9), 1608–1624.

Mkono, M. & Holder, A. (2019). The future of animals in tourism recreation: Social media as spaces of collective moral reflexivity. *Tourism Management Perspectives*, *29*, 1–8.

Von Essen, E., Lindsjö, J., & Fouache, A. S. C. (2020). *INSTAGRANIMAL–Animal Welfare and Ethical Challenges of Animal-Based Tourism*. Retrieved from https://portal.research.lu.se/portal/files/77882057/report_instagranimal_aug_27_28_2019.pdf

Wearing, S. & Jobberns, C. (2015). From 'Free Willy' to 'Sea World': Has Ecotourism improved the rights of Whales. *Animals and Tourism: Understanding Diverse Relationships*, 67, 75.

David A. Fennell

15 Afterword: On tourism, animals, and suffering – lessons from Aeschylus' *Oresteia*

Introduction

In 1999, I read with interest a paper by Holland, Ditton, and Graefe (1998) that linked billfishing (angling for marlin, swordfish, and sailfish) with ecotourism. The authors argued that billfishing could be a legitimate form of ecotourism because it satisfied several key criteria. It (1) occurs in a unique natural resource, (2) attracts a unique clientele, (3) is a responsible use of natural resources, (4) provides economic support for resource conservation, and has economic advantages over other uses, and (5) directs economic assistance to the local economy. The 1990s was a fruitful time for ecotourism, because there was the belief that just about anything could qualify as ecotourism if it hit on any number of criteria such as the ones identified above.

The Holland et al. paper prompted me to question not their scholarly approach, but rather the whole enterprise and legitimacy of using animals for ecotourism purposes (Fennell, 2000a; Fennell, 2000b). Something seemed intuitively wrong with angling as ecotourism, especially the pursuit of billfish, which include some of the most impressive beasts in the marine world. I argued that in billfishing there is the intention to pursue an animal, mindful that such pursuit would lead to pain and sometimes death of the animal for purposes of recreation; fishing was felt to be more consumptive than other forms of ecotourism (e.g., birdwatching); and that ecotourists and sport/recreation anglers do not share the same values. I say *intuitively*, because at the dawn of the new millennium there was only a blank slate. No scholarly work at hand in our tourism journals and books to answer questions that I only partially articulated at the time: What about the interests of animals used in ecotourism, and tourism more generally? What are the consequences to individual animals, if we focus only on the health of populations and ecosystems? Should ecotourism promote suffering?

Thankfully, the questions that were confounding me back in 2000 are now a focus of regular, intensive study in the tourism field two decades later. We are presently much more conversant on animal rights, animal welfare, utilitarianism, ecofeminism, ecocentrism, and more recently, issues around contractarianism and justice (Fennell & Sheppard, 2020). Not surprisingly, we've taken our lead from animal ethics scholars from outside of our field (e.g., Singer, 1975; Regan, 2004), like so many other themes in tourism studies research. And we understand from the broader animal ethics literature that, for example, ecocentrists and animal liberationists are miles apart in their views on who/what is deserving of moral consideration in the natural world (Sagoff, 1984).

https://doi.org/10.1515/9783110664058-015

We can safely say that Holland et al. (1998) were being slavish to an ecocentric way of looking at the world in their views on billfishing, i.e., we are on safe ground if, morally speaking, the stability and integrity of the natural world is given primary consideration.

My purpose here is to add further weight to the burgeoning thought on tourism, animals, and ethics – to expand our working knowledge – as we ponder questions around if we should use animals in tourism as workers, why we use them, and how. I enlist Aeschylus' compelling Greek tragedy, *Oresteia,* to articulate meaning around several key themes: conflicting narratives around justice, invengeance, pathei mathos, the sacrificial economy, and especially suffering. In this task, I draw liberally on Stefan Dolgert's (2012) interpretation of these concepts in his comprehensive views on justice, politics, democracy, and Greek tragedy.

Suffering holds an important place in the history and philosophy of animal ethics. This is not the place to explore it in great depth, but a couple of examples will hopefully suffice. It was Jeremy Bentham (1780/1982) who adopted, not a perspective based on rationality, language, or personhood when it comes to the moral considerability of animals, but rather one based on sentience. As a utilitarian scholar, Bentham was concerned about the balance of pleasure and pain, and the satisfaction of preferences and interests. Bentham famously argued that "The question is not, Can they reason?, nor, Can they talk? but, Can they suffer?" Nearly 200 years later, the Oxford scholar Richard Ryder, widely viewed as one of the key founders of contemporary animal ethics, produced a leaflet that conceptualized suffering around negative experiences such as fear, grief, anxiety, distress, and the effects of captivity. As the principal architect of speciesism, Ryder (1970, n.p.) wrote:

> Quite apart from the right to live, one clear moral criterion is suffering, the suffering of imprisonment, fear and boredom as well as physical pain. If we assume that suffering is a function of the nervous system then it is illogical to argue that other animals do not suffer in a similar way to ourselves – it is precisely because some other animals have nervous systems so like our own that they are so extensively studied [. . .] If we believe it is wrong to inflict suffering upon innocent human animals then it is only logical, phylogenically-speaking, to extend our concern about elementary rights to the nonhuman animals as well.

While it is not the purpose of this volume to chronicle the suffering of animals within the tourism industry, nevertheless it is impossible to consider working animals in tourism without it arising. Indeed, several authors in this volume touch on the concept of suffering. Quintero Venegas and López López (Chapter 4) oppose the fact that the use of animals, like "zonkeys," in trade and consumption is justified in the absence of *extreme* suffering; they lament that lesser degrees of suffering are deemed acceptable by the industry as an accepted norm of "business-as-usual." Daspher (Chapter 2) consults work by Coulter (2016) and Porcher (2011) to illustrate that both human and non-human suffering is linked in the workplace. Tully and Carr (Chapter 6) investigate farm attraction websites in New Zealand and find that only one of 29 websites discussed the reduction of suffering. Tourist satisfaction at these sites appears to be more important than the suffering of animals. On bullfighting in Mexico, López López and Quintero Venegas (Chapter 5) similarly demonstrate that bulls are made to endure significant psychological and physiological suffering before they are allowed to enter the bullring, for the purpose of suppressing the full strength of bulls. Vasilopoulou's (Chapter 10) case study on the donkeys of Santorini reveals suffering of donkeys who perform difficult transportation work during long, hot days of Greek summers, while also shedding light on the complex relationships between muleteers and their animals (in an unequitable tourism economy). Finally, humans working in monkey sanctuaries in Costa Rica conveyed what Speiran (Chapter 11) argues to be maternal instinct, and as a result, releasing abused monkeys that have been made to suffer back into the wild is richly rewarding for them.

Aeschylus' *Oresteia*

The *Oresteia* is one of the finest examples of dramatic poetry in existence and the only complete Ancient Greek trilogy that survived to the present day. It is comprised of three plays, *Agamemnon, Libation Bearers*, and *Eumenides*, which are connected both in subject matter and plot. It was first performed in 458 BCE, at the Festival of Dionysus in Athens, and takes aim at the problem of violence and justice in Greek society and how best to respond to it (Dugdale & Gerstbauer, 2017). Key to the structure of *Oresteia* is a move away from retaliatory justice, to justice based on reason and law.

In the first of the three plays, Agamemnon returns home to Argos after 10 years conquering the city of Troy with the Tojan princess, Cassandra, as his war concubine (one of the "spoils of war"). Although initially and enthusiastically welcomed home by Agamemnon's wife, and queen, Clytemnestra, she later murders Cassandra and Agamemnon because of his unfaithfulness; his killing of their daughter Iphigenia to appease the gods over a previous transgression; and to more formally embrace her own new lover, Aegisthus. The dead bodies of Agamemnon and Cassandra are publicly displayed in the streets of Argos as Clytemnestra's just revenge.

Libation Bearers, Aeschylus' second play, takes place several years later at the palace of Argos. The son of Agamemnon and Clytemnestra, Orestes, kills Clytemnestra and Aegisthus to avenge the death of Agamemnon in a retaliatory fashion, which was the norm in Greek society. Mirroring the first play, Orestes also places the dead bodies in the streets of Argos for public scrutiny as an equivalent crime and punishment.

In the third play, *Eumenides*, Orestes is given sanctuary from Apollo at Delphi, but later tried in Athens by court and jury involving 12 honorable Athenians – the first time in Greek society that justice was configured in such a way. Orestes is acquitted of the charges of murder, but the Furies (also called Erinyes and Eumenides), three female beast-like goddesses that seek vengeance for those who have committed evil acts against those unable to avenge themselves, threaten the stability of Athens.

Athena tames and placates the Furies by giving them a new role as protectors of justice in Athens, with the agreement they can still unbridle their wrath on those who deliberately unbalance civil harmony.

Sacrifice, suffering, and the *Oresteia*

Writing on the relationship between animals, sacrifice, and justice from the perspective of Aeschylus' trilogy, Dolgert (2012) argues that the regular sacrifice of animals in Greek society was about finding suitable surrogates to honor conflicting narratives that we see in the *Oresteia*. The emergence of a sustained sacrificial economy was premised on two main themes: invengeance and the pathei mathos .

First, animals were used in sacrifice because they are unable to speak for themselves in challenging or avenging what passes as a criminal death. Dolgert (2012) has termed this *invengeance* or "the inability [. . .] of the victim-group to be avenged" (p. 270). Being non-human, automatically categorizes one as invengeant, which in turn means that they can be sacrificed without the burden of guilt, and without the chance of enduring the vehemence of avengers, as we saw in the case of Orestes and the Furies, above. The killing of animals does not sustain the retaliatory cycle of vengeance that is such a prominent theme in the Oresteia. This is because these killings are non-criminal deaths legitimized by a sacrificial economy that is deeply embedded as a political and cultural practice (Derrida, 1991).

Second, because animals are animate and possess sentiency, they are a necessary component of the relationship between tragedy and democracy. Dolgert argues that Greek democracy drew form and function from suffering, which became a vital source of political knowledge. This is termed the *pathei mathos*, and translates as learning or knowledge through suffering in achieving political wisdom (Dolgert, 2012; Sommerstein, 2010). There is thus a civic function of tragedy played out in the spectacle that was Ancient Greece, which depended on the killing of animals that were conceived as "like" human in imparting a sense of horror in the eyes and minds of the audience (see Euben, 1990). Dolgert observes that this translates into a fetishization of animal death that shapes social and political practice. Taken one step further, the community feeds off the violence that is inherent in sacrifice as a pressure-release in preventing civil conflict that may emerge after too much pent-up aggression (see Girard, 1979). (Bullfighting, as discussed in Chapter 5, may be an apt representation of this fetishization of animal death and the feeding off of violence.) As such, the scapegoat – the animal – is merely the symbol of this civic strife, and the sacrificial animal ultimately becomes the community's savior (Dolgert, 2012). The scapegoat dies so the community might thrive.

Dolgert (2012) concludes his work by asking if we wish to know where the boundaries of the ethical community lie, then we should be cognizant of the political subconscious that shapes our ideas of community. When communities anesthetize suffering and pain – or forget suffering as Chiurazzi (2012) observes – such becomes a strategy for domination. It then becomes rational to mute and sedate animal suffering in practices such as animal experimentation, factory farming, and the taken-for-granted work that animals perform in tourism, so we can get on with the business of satisfying our interests and feeling good about ourselves. But shouldn't suffering have humanizing and moralizing effects instead of domination and the elevation of self-interest (see Lombard, 1984)? Indeed, *Oresteia* is as much about moral responsibility as it is about the other themes discussed above.

Discussion and conclusion

What does the story of the *Oresteia* provide for us that will have relevance for tourism practice now and in the future? *Oresteia* is conceived as a progression from barbarism to civilization through the advent of a new and more progressive form of justice. It tackles the thorny issue of invengeance under the belief that it is only through suffering that democratic societies can draw knowledge about conflicting narratives in achieving wisdom (pathei mathos). Animals as animate and sentient beings conveniently fill the role of sacrificial lamb in a sacrificial economy for the purpose of restoring the community back to a position of stasis. And it is about the suppression of pain and suffering so we can live emotionally and cognitively free and unencumbered.

Tourism's sacrificial economy involves a dizzying array of animal uses that have ultimate (death) and enduring (sustained suffering) costs to animals in the name of the tourism good. Factory farming of food, hunting, fishing, and animals being worked to death in tourism are prime examples of the former. Enduring costs include animal lives that are instrumentalized and marginalized through a range of activities, like the many discussed here in this book.

In much of this use we get lost in the commerce and pleasure that hides us from the consequences of our actions. Tourists elevate their actions as heroic through simple social exchanges: money for experience, i.e., we have done our job as "good" tourists by putting money in the hands of needy service providers. It's the Great Chain of Being at work. But instead of God and the angels at the summit, it's now commerce, commodification, and neoliberalism that sit atop the hierarchy, followed by tourists and operators, and animal others positioned no better than during the time of Aristotle. We maintain this Cartesian mode of thinking and acting because our contractarianist way of fabricating the world tells us that fair exchanges can only take place between rational agents who are able to reciprocate.

Since animals can neither rationalize nor reciprocate, as conventional wisdom tells us, secondary moral status is the most we can ever grant them (see Rawls, 1971).

Despite the rather sobering account of the state of affairs illustrated above, there is room for optimism. Presently leadership is coming from a growing faction of scholars who are making animal ethics a bona fide sub-field in tourism studies. Change is also coming from industry. Service providers who abuse animals are now routinely rejected by some of the biggest players in the industry. The Intrepid Group is a case in point, who in 2014 was the first to ban elephant rides because of the suffering that these animals have to endure in working for the tourism industry (https://www.intrepidtravel.com/adventures/change-elephant-rides/)

More momentous strides will be taken if organizations like the UNWTO, our global champion for responsible and sustainable tourism, decides to step up to the table. Adopting the proposed Article 11 on the welfare considerations of animals used in tourism (Fennell, 2014) would signal to the world the essential role that animals play in the fabric of global tourism. Section 7 of the Article observes that, "All individuals or groups that use animals for tourism purposes should have access to organisations that have as their mandate the reduction of animal suffering through the implementation of proper welfare standards and care appropriate for all species used." (Fennell, 2014, p. 992). It should not be the reduction of animal suffering that is the goal, but rather the *elimination* that must be at the heart of a truly responsible tourism industry.

The *Oresteia* was built around the progression of justice, from a traditional model based on retaliation, to one based on law and reason. Justice for animals used in tourism must also evolve as a force for change. Just and unjust uses of animals can now be "weighed" according to a "scales of justice" framework based on the amount of respect, dignity, welfare, and suffering that individual animals experience as workers, entertainers and performers in tourism (Fennell & Sheppard, 2020). This includes the identification of types of justice such as "deep," "intermediate," "shallow," and "no justice" forms conceived from conventional normative ethics (i.e., what we are told to do) and virtue ethics (i.e., what we ought to do as individuals). There can be "no justice" for animals in bullfighting because of the outright denial of respect, dignity, and well-being. By contrast, we can have "deep justice" if animals are allowed to live free and not subject to human control and manipulation. In the *Oresteia*, it was the Furies who ultimately emerged as protectors of justice. Who will step forward in tourism to promote animal justice and eliminate suffering, is a topic of considerable importance.

The tragedy is not only in the play, *Oresteia*, which imparts a valuable lesson. It is also found deeply rooted in tourism ignorance and arrogance. How we pull ourselves out of this quagmire will be one of the great challenges of our future. Indeed, 20 years on, we are in a much better position to answer questions on the moral acceptability of billfishing as ecotourism. But it is now much more than ecotourism. This volume on working animals in tourism is a further testament to how the tourism and animal ethics sub-field has matured over a relatively short period of time. A focus on animal workers in tourism is quite simply another gap that must be filled if knowledge is to inform practice.

We have the opportunity – the obligation – to reshape tourism scholarship and practice through deep, lateral, sustained, and transdisciplinary thinking, which accentuates the intrinsic value of animals and how we should meaningfully coexist. In engaging this end, we will have firmly established that suffering is an artifact of the past in delineating responsible *touranimalscapes* of the future – evolving patterns of animal concern and use (including refusal of use), on the part of individuals, cultures, regions, and organizations connected with the tourism industry.

References

Bentham, J. [1780] (1982). *An Introduction to the Principles of Morals and Legislation*, edited by J.H. Burns and H.L.A. Hart, London: Methuen, 1982 (Athlone Press 1970).

Chiurazzi, G. (2012). Pathei mathos: The political-cognitive value of suffering. In J. Malpas and N. Lickiss (Eds.), *Perspectives on human suffering* (pp. 23–32). New York: Springer.

Coulter, K. (2016). *Animals, work and the promise of interspecies solidarity*. New York: Palgrave Macmillan.

Derrida, J. (1991). Eating Well. In E. Cadava, P. Connor, & J-L. Nancy (Eds.*) Who Comes after the Subject*. London: Routledge.

Dolgert, S. (2012). Sacrificing justice: suffering animals, the Oresteia, and the masks of consent. *Political Theory*, *40*(3), 263–289.

Dugdale, E. & Gerstbauer, L. (2017). Forms of justice in Aeschylus' *Eumenides. Polis, The Journal of Ancient Greek and Roman Political Thought*, *34*, 226–250.

Euben, P. (1990). *The tragedy of political theory*. Princeton, NJ: Princeton University Press.

Fennell, D. A. & Sheppard, V. (2020): Tourism, animals and the scales of justice. *Journal of Sustainable Tourism*, DOI: 10.1080/09669582.2020.1768263

Fennell, D. A. (2014). Exploring the boundaries of a new moral order for tourism's global code of ethics: an opinion piece on the position of animals in the tourism industry. *Journal of Sustainable Tourism*, *22* (7), 983–996, DOI: 10.1080/09669582.2014.918137

Fennell, D. A. (2000a) Comment: ecotourism on trial-the case of billfish angling as ecotourism, *Journal of Sustainable Tourism*, *8* (4), 341–345, DOI: 10.1080/09669580008667370

Fennell, D.A. (2000b). Tourism and applied ethics. *Tourism Recreation Research*, *25(1)*, 59–70.

Frey, R.G. (1980). Interests and Rights: The Case Against Animals. Oxford. The Claredon Press.

Girard, R. (1979). *Violence and the sacred*. Baltimore, MD: Johns Hopkins University Press.

Holland, S.M., Ditton, R.B. and Graefe, A.R. (1998). An ecotourism perspective on billfish fisheries. Journal of Sustainable Tourism 6 (2), 97–116.

Lombard, D. B. (1984). Pathei mathos in three tragedies of Euripides. PhD dissertation, Rand Afrikaans University, Johannesburg, South Africa. Retrieved at.

Mendl, M. & Paul, E. (2004). Consciousness, emotion and animal welfare: insights from cognitive science. *Animal Welfare*, *13*(1), 17–25.

Porcher, J. (2011). The relationship between workers and animals in the pork industry: A shared suffering. *Journal of Agricultural and Environmental Ethics*, *24*(1), 3–17.

Rawls, J. (1971). *A theory of justice*. Oxford: Oxford University *Press*.

Regan, T. (2004). *The case for animal rights*. Berkeley. University of California Press.
Ryder, R. (1970). Speciesism. Privately printed leaflet, Oxford University. Retrieved at https://web.archive.org/web/20121114004403/http://www.criticalsocietyjournal.org.uk/Archives_files/1.%20Speciesism%20Again.pdf
Sagoff, M. (1984). Animal liberation and environmental ethics: bad marriage, quick divorce. *Osgoode Hall Law Journal 22*, 297–307.
Singer, P. (1975). *Animal liberation*. New York. New York Review/Random House.
Sommerstein, A. H. (2010). The tangled ways of Zeus: And other studies in and around Greek tragedy. Oxford: Oxford University Press.

List of contributors

Mary Benbow is an Associate Professor in the Department of Environment and Geography at the University of Manitoba. She primarily focuses her research upon the social, cultural, and environmental implications of zoos and aquariums. Her research looks at how these cultural institutions affect human perceptions of animals and conservation that in turn impact how animals are managed. This forms part of a broader field animal geographies that looks at the multifaceted roles that animals play in our lives and seeks to illuminate the complex relationships between humans and animals.

Jill Bueddefeld is currently a post-doctoral fellow in the Faculty of Kinesiology, Sport, and Recreation at the University of Alberta. Her research focuses on nature-based tourism, visitor experiences, environmental education, with an emphasis on free-choice learning and sustainable behavior change. Jill has worked on an array of free-choice and transformative learning projects, which have explored different contexts such as festivals, agritourism, zoos, and parks and protected areas.

Anita Burkevica is Zoo Director and Doctor of Veterinary Medicine at Parken Zoo in Sweden. She was born in Latvia; growing up she was always surrounded by animals. Her interest in wildlife started when she turned 17 and worked as a zookeeper in Riga Zoo. After her 18th birthday, she started working at a small animal hospital and applied to study veterinary medicine at the University of Agriculture in Jelgava where she focused on wild, exotic animal and conservation medicine. She had done several internships at Kolmården Zoo, Sweden, where she also started working after her graduation in 2014, while also working at a small animal hospital. In 2015, she started as veterinarian in Parken Zoo, Sweden, and two years later, she received the offer to become Zoo Director to improve animal welfare at the zoo.

Neil Carr is a Professor in the Department of Tourism at the University of Otago. His research focuses on understanding behavior, welfare, and rights within tourism and leisure experiences, with a particular emphasis on animals, children and families, and sex.

Erika Andersson Cederholm is Associate Professor of Sociology at the Department of Service Management and Service Studies, Lund University in Sweden. Her research embraces areas such as rural lifestyle entrepreneurship, tourism, service work, the meeting- and event industry, and creative labor. Theoretically, she works in the fields of economic sociology, cultural sociology, and working life studies, with an interest in the intersections between professional, commercial, and personal spheres of life, such as friendship and gift exchanges in networks and in workplaces. Her recent projects are on hunting tourism entrepreneurship, and on the working conditions in the digital games industry.

Katherine Dashper is Reader in the School of Events, Tourism, and Hospitality Management at Leeds Beckett University, UK. Her research focuses predominantly on human-animal relationships in the context of tourism, events, and leisure, with particular interest in horses. She has published widely in international journals and books on issues related to human-animal relationships, gender, tourism, and events management. She is author *of Human-animal relationships in equestrian sport and leisure* (Routledge, 2017), and co-editor of four books on tourism, events, and sport.

https://doi.org/10.1515/9783110664058-016

David A. Fennell is a Professor in the Department of Geography and Tourism Studies at Brock University. He researches mainly in the areas of ecotourism, tourism ethics, and moral issues tied to the use of animals in the tourism industry and sustainability. A major thrust of his research involves the use of theory from other disciplines (e.g., biology, philosophy) to gain traction on many of tourism's most persistent issues and problems. Fennell is the founding Editor-in-Chief of the *Journal of Ecotourism* and is an active member on editorial boards of many academic journals.

Adélaïde Fouache is a student in the Environmental Communication and Management Master program at SLU, Uppsala (Sweden). She wrote her Master's thesis on the impacts of the touristic activity that is whale watching on the endangered Southern Residents Killer Whales living in the Salish Sea. She would like to deepen her knowledge on marine tourism and marine wildlife by working with an NGO on communicating the consequences of human activities on the surrounding wildlife and the actions that can be taken. She has a particular interest in endangered marine mammals.

José-Carlos García-Rosell is a Senior Lecturer in Responsible Tourism Business at the Multidimensional Tourism Institute (MTI), Faculty of Social Sciences, University of Lapland, Finland, and Adjunct Professor in Tourism at the Faculty of Tourism, University of Maribor, Slovenia. He works in the fields of corporate social responsibility (CSR), business ethics, and responsible management education in tourism. His current research agenda comprises four areas: human-animal relations, sharing economy, responsible tourism experiences, and experiential learning. His work has been published in various books and journals such as *Journal of Business Ethics*, *Management Learning*, *Journal of Sustainable Tourism*, *Tourism Geographies*, and *Tourism Recreation Research*.

Yulei Guo is a researcher in Chengdu Research Base of Giant Panda Breeding. Her current research concerns wildlife-human interactions in tourism, environmental aesthetics, conservation, and cuteness studies. Her recent project, "Pink Sustainability," prioritizes the aesthetics of cuteness in tourism consumption based on investigations of giant panda's global images and perceptions. She is interested in understanding the use of cute images in sustainable development and conservation (e.g. WWF's logo), as cuteness has become an internationally recognized aesthetic category. The project Pink Sustainability also pays attention to women's roles in participating and producing discourses of sustainability through consuming cuteness images.

Nigel Halpern is Professor of Air Transport and Tourism Management with the Department of Marketing at Kristiania University College in Oslo, Norway. He has previously worked at Molde University College, Norway; London Metropolitan University, UK; the UK Department for Transport, Local Government and the Regions; the UK Civil Aviation Authority; and PGL Travel in the UK, France, and Spain. Nigel teaches and conducts research and consultancy in air transport and tourism. His main interests are in accessibility, digital transformation, marketing and strategy, service quality, interorganisational relations, and geographies of air transport and tourism impacts.

Marcus Hansen is a lecturer in Hospitality, Tourism, and Events at Wrexham Glyndwr University. He holds a PhD in tourism management from Manchester Metropolitan University (MMU), which explored stakeholder collaboration in adventure tourism. Marcus teaches and conducts research in relation to tourism and event management. His research interests are particularly around the topics of adventure tourism and accessible tourism and events, including traveling with a vision impairment and making destinations dementia-friendly.

Hindertje (Hin) Hoarau-Heemstra is currently employed as associate professor at the Nord University Business School in Bodø, Norway. Her research interests include management, sustainability, innovation, and collaboration. She has mainly done research in the context of tourism, focusing on Nordic nature-based tourism and animal-based tourism experiences.

Michael Jones is an advisor to the Swedish Biodiversity Centre. He teaches systems thinking for sustainable development to undergraduate and Master's degree students and does policy-oriented research for the management of ecosystems as social-ecological systems. He has 47 years of experience in conservation and development, mostly in southern and eastern Africa and the western USA. Mike's background is in wildlife ecology, rangeland ecology, protected area management, community conservation, and the application of systems science to ecosystem management approaches such as nature-based solutions. He is Chair of the Resilience Theme Group in IUCN's Commission on Ecosystem Management; and a member of the International Association of Impact Assessors.

Carol Kline is an Associate Professor of Hospitality and Tourism Management at Appalachian State University. Her teaching and research interests have historically focused broadly on tourism sustainability, including topics such as foodie segmentation, craft beverages, agritourism, tourism entrepreneurship, and tourism in developing economies. However, she now gears her research solely on animals, and she teaches a course called Animals, Tourism, & Sustainability. She is part of the Race, Ethnicity, and Social Equity in Tourism (RESET) initiative, which includes animals within the study of social equity. She is founder of Fanimal Inc., a non-profit that helps individuals find animal-focused careers.

Johan Lindsjö is a veterinarian and graduated 1995 at the Swedish University of Agricultural Sciences (SLU). As a clinician he specialized in diseases of dogs and cats, he has a MSc in wildlife health from the Western College of Veterinary Medicine, University of Saskatchewan, Canada, and he has worked in several international veterinary volunteer projects. He is doing a residency in Animal Welfare, Ethics and Law (ECAWBM). Johan is a qualified officer at the Swedish Centre for Animal Welfare and a lecturer at the Department of Animal Environment and Health, SLU. He is particularly interested in the 3Rs in a broader perspective and One Welfare.

Álvaro López-López has a PhD in Geography from the Arts Faculty of the National Autonomous University of Mexico (UNAM). He is currently a senior researcher at the Institute of Geography, UNAM, and also a professor in the undergraduate and graduate programs in Geography at UNAM where he teaches "Space and tourism." His basic line of research has been the geography of tourism in Mexico, from which he has developed research projects about sex tourism, dark tourism, and the intersection between the geography of tourism and the geography of animals, among others. He was the President of the Mexican Academy of Tourism Research.

Brent Lovelock is a Professor with the Department of Tourism at the University of Otago. He is a social scientist with a passion for advancing sustainable tourism through his research and teaching. He is particularly interested in nature-based tourism and how we manage the experience for the benefit of all stakeholders, including nature herself. He is the co-author of *The Ethics of Tourism*, a volume which challenges us to consider more ethical approaches to managing tourism. He is part of a university team that has undertaken a seven-year program of research on education for sustainability.

Mariana Macieira Reis is a veterinary medicine student at ICBAS (Institute of Biomedical Sciences Abel Salazar) of the University of Porto, Portugal. She is an animal and nature enthusiast and as such chose to complete her curricular internship in pathology and infectious diseases of wildlife at the SVA- National Veterinary Institute of Sweden. This resulted in a master's thesis regarding the health status of the Swedish muskrat population, focusing on *tularemia*. Previously, she has participated in her student's association as the coordinator of the veterinary department and president of the General Assembly, as well as a volunteering project working with animal shelters.

Scott McCabe is Professor of Marketing and Tourism at the Nottingham University Business School. Scott's research has focused on aspects of the psychology and social psychology of tourist consumer behavior and decision-making, particularly pro-social and pro-environmental tourist consumption. Scott's research focuses on qualities of tourist experiences and how they are designed and marketed. His work on social tourism is recognized internationally, and he has worked alongside European colleagues and beyond to make the case for social inclusion through tourism. He is also an elected Fellow of the International Academy for the Study of Tourism and is the current co-Editor in Chief of *Annals of Tourism Research*.

Nadezda Nazarova is currently employed as associate professor at the Nord University Business School in Bodø, Norway. With the background in accounting and management control, her research interests have covered the role of non-human actors in organizations in the context of logistics, risk management, and smart societies in the High North.

Gino Jafet Quintero Venegas is an associate researcher at the Instituto de Investigaciones Sociales of the National Autonomous University of Mexico (UNAM) within the research line "social space." His research interests are tourism and space, animal geography, and animal ethics. His formation is Geography in undergraduate, Master's and PhD levels at UNAM. He also completed a postdoctoral fellowship at the UNAM's Program of Bioethics (PUB) with the project "Bioethics, Geography and Zootourism. Ethical reflections on the use of non-human animals as tourist resources and cultural elements." Currently, he is Professor of "Animal Geography" and "Geography and Ethics" at the Arts Faculty at UNAM.

Jillian M. Rickly is Professor of Tourism at the University of Nottingham and the Series Editor for De Gruyter Studies in Tourism. She is a tourism geographer with research interests in authenticity/ alienation in tourism, accessibility and tourism mobilities, and critical animal studies in tourism and recreation. Her work has been published in the *Annals of Tourism Research, Journal of Sustainable Tourism, Tourist Studies, Cultural Geographies,* and *Mobilities*, among others. She is co-author of *Tourism, Performance, and Place: A Geographic Perspective* (2015, Ashgate) and *Risky Business: Management Challenges for the Adventure Tourism Industry* (forthcoming 2021, De Gruyter), as well as co-editor of numerous books on mobilities, events, and authenticity in tourism studies.

Siobhan I. Speiran is a PhD candidate in Environmental Studies at Queen's University, and a part of The Lives of Animals Research Group. Generously funded by a SSHRC Bombardier Scholarship, she focuses on the lives of monkeys in Costa Rican sanctuaries. Interdisciplinary in concept and approach, her work considers how sanctuaries–as sites of ecotourism–contribute to the conservation and welfare of monkeys. Drawing on scholarship from welfare and conservation biology, animal geography, ethnoprimatology, and tourism studies; her approach is bottom-up, working closely with sanctuary owners and networking outwards and upwards to understand structural hierarchies of primate management in Costa Rica.

Linda Tallberg is Assistant Professor in the Department of Management and Organization at Hanken School of Economics in Finland. She has previously held academic positions at Griffith University (Australia) and Åbo Akademi University (Finland). Her research focuses around animals in organizations, dirty work, animal shelters, emotions, critical animal studies, and alternative research methods such as crystallization. She is editor of *TRACE: Journal of Human Animal Studies* and on the board of the *Finnish Society for Human Animal Studies*.

Lara Tickle received her doctorate from the Swedish University of Agricultural Sciences (SLU), Division of Environmental Communication in Uppsala, Sweden, where she most recently worked on the project "Challenges Facing Swedish Hunting Ethics." Her work is centered on modern processes of technological advancement and demographic trends that affect ethical relations to nature and wildlife. Hunting is often at the center of her research as a platform for human-wildlife interactions. She is experienced in governmental policy and has a background in physical and human geography.

Paul A. G. Tully holds a Bachelor's in Tourism Management from the University of Central Lancashire, UK, and a Master's in Sociology & Social Research from Newcastle University, UK. He has research interests that span the critical study of tourism in the social world and has a current focus on human-animal relationships. His current projects concentrate on the use and treatment of domesticated animals during the human experience. In particular, he aims to consider with an animal-centric lens the role of farm animals in tourism.

Christina Vasilopoulou holds a master's degree in Sociology from Panteion University, Greece, where she completed her master dissertation on "Animals and biopolitical power in the Attica Zoological Park." She has also a bachelor's degree in Economics and a degree in Education and Didactics from the Athens University of Economics and Business. Also, she is a member of the scientific research team of the AUEB's Teacher Education Program. She works as a private high school teacher in economics, sociology, and business administration. Her main research interests are the relationships between human and non-human animals, concerning matters of rights, ethics, and relativity.

Erica von Essen is an interdisciplinary human-animal studies researcher with a PhD in Environmental Communication from the Swedish University of Agricultural Sciences. Currently at the Norwegian Institute for Nature Research, von Essen's fields of expertise include changing relations with wildlife in modernity, illegal hunting, rural sociology, and animal ethics – both descriptive and normative. She is published across criminology, sociology, ethics, and wildlife management journals and is a regular reviewer for more than 25 journals. Von Essen is committed to exploring the new roles that we carve out for non-human animals in society and to highlight the ethics of these relations.

List of figures

https://doi.org/10.1515/9783110664058-017

Index

https://doi.org/10.1515/9783110664058-018